The Strength-Ductility Paradox

David J. Fisher

Published by **Materials Research Forum LLC**
Millersville, PA 17551, USA

Published as part of the book series
Materials Research Foundations
Volume 168 (2024)
ISSN 2471-8890 (Print)
ISSN 2471-8904 (Online)

Print ISBN 978-1-64490-322-3
ePDF ISBN 978-1-64490-323-0

Distributed worldwide by

Materials Research Forum LLC
105 Springdale Lane
Millersville, PA 17551
USA
http://www.mrforum.com

Printed in the United States of America
10 9 8 7 6 5 4 3 2 1

Table of Contents

Introduction

The inverse relationship between strength and toughness plagued engineers for the centuries before it was even known that metals had an internal structure. It continued to plague metallurgists even when metallography developed at the turn of the 20th century and when the role played by dislocations became apparent during the following decades. The essential problem of toughness is, of course, a want of ductility. The strength-ductility paradox is however rather more subtle, and was spotlighted barely two decades ago[1]. The seminal paper referred specifically to the improved plasticity in titanium, subjected to high-pressure torsion, which resulted from ever more severe deformation while the strength remained the same. Indeed, it is also sometimes referred to as the SPD-paradox, where SPD is the initialism of *severe plastic deformation*. Subsequent work[2] revealed that short-term annealing of the deformed material increased the strength and ductility simultaneously. Annealing at higher temperatures led to additional increases in ductility and strength. The same phenomenon was later observed in other pure metals, in alloys and in metal-matrix composites. Severe plastic deformation, usually via equal channel angular pressing and high-pressure torsion, now permits the preparation of materials possessing unique properties. The phenomenon is still however lacking a definitive and universal explanation, especially for pure metals. Various ideas have been advanced, such as the almost venerable one of so-called non-equilibrium grain boundaries possessing increased energy due to dislocation-absorption during severe plastic deformation that then imparts an increased diffusivity. This would imply that grain-boundary sliding is implicated, in the case of ultrafine-grained materials, as being a deformation mechanism which can provide adequate ductility. Another plausible explanation for increased strength plus ductility in ultrafine-grained material is rapid diffusion along grain boundaries. A third explanation pertains to metals which possess an heterogeneous microstructure which comprises large grains surrounded by small ones or small grains surrounded large ones. The large grains would then provide good ductility and the small grains would provide a high strength.

The following provides an *aperçu* of the incidence of the strength-ductility paradox in a wide range of materials, and the range of mechanisms involved.

Pure Metals

In the seminal paper on the subject it was pointed out that the yield strength of aluminium and copper increased monotonically, while the elongation-to-failure decreased. The same tendency was also true of other metals and alloys. It was further noted that an anomalous combination of high strength and high ductility was produced by severe plastic deformation, and this was attributed to the particular nanostructures generated by such processing: a combination of ultra-fine grain size and high densities of dislocations was thought to facilitate deformation via new mechanisms. Experimental tests and computer-modelling of bulk pure metals which had been subjected to severe plastic deformation were later used[3] to explore the phenomenon at various scales. The grain size of copper could be refined to 150 to 200nm, and that of aluminium could be refined to 1μm. The grain size in alloys could be made several times smaller than it could in pure metals. The effect of processing parameters upon the flow of the deformed material, and the elastic stress-fields which existed, were studied at the macro-level. Grain and sub-grain size-changes were studied at the meso-level. The micro-level computer-simulations predicted the densities of extrinsic grain-boundary dislocations, and the effect of the deformation upon the extent of static and dynamic atomic displacements. This made it possible to explain the simultaneous observation of high strength and ductility, increased diffusion and the low-temperature and/or high-rate superplasticity of severely deformed bulk nanomaterials. These can be characterized as having a grain size of tens to hundreds of nanometres, meaning that the volume fraction of boundary area can even exceed the volume fraction of the bulk of the grain. Given that grain boundaries have a structure which is very different to that of the bulk, they become the factor of interest because of their highly non-equilibrium state and high defect-content; especially of extrinsic grain boundary dislocations. These generate long-range elastic stresses which change the interatomic distances and thus the lattice parameter, the elastic modulus, the Debye temperature, the linear thermal expansion coefficient and the diffusion coefficient. It is therefore natural that nanostructured materials should exhibit considerably different structural properties to those of coarse-grained ones. In practice, high-pressure torsion and equal-channel angular pressing are the main severe-deformation techniques and, in the former case, a billet is subjected to a monotonic shear deformation caused by frictional forces which act between the billet and the heads, during twisting of one of the heads. The evolution of the structure can be monitored at various stages, and this technique produces a considerably smaller grain size than does equal channel angular pressing.

In a sweeping review of the subject a decade ago, it was noted[4] that the grain size is generally the predominant structural feature of polycrystalline metals. At temperatures of up to 50% of the absolute melting point - the accepted rule-of-thumb criterion for the onset of hot-working - the yield stress varies according to the Hall-Petch law. Because of its inverse nature, a reduction in grain-size naturally leads to an increase in the overall strength. Conventional industrial thermomechanical processes are however incapable of reducing the grain size down to the micrometre-scale at which high strengths, and the paradox, become so noticeable. Whence the interest in the extreme plastic deformation of bulk coarse-grained solids. It was also noted that the popular depiction (e.g. figure 1) of the strength-ductility paradox is only qualitative, in that there is no definitive relationship for the envelope which confines conventional materials and separates them from high-strength high-ductility materials. The elongation-to-failure, for example, can depend upon the specific design of the test-piece used. It was therefore proposed that normalized (dimensionless) coordinates should be used in which the yield stress for ultrafine-grained samples was divided by the yield stress for coarse-grained samples, and similarly for the elongation to failure. This type of plot (not shown) then had horizontal and vertical demarcation-lines through the points where the ratios were equal to unity. Experimental points to the left and/or below the lines then corresponded to conventional behavior where either the strength or ductility was worse than that of coarse-grained material. Points in the top right-hand corner were then in the high-strength and high-ductility region where the strength-ductility paradox was evaded. This dimensionless plot was not however generally adopted. It was further noted that ultrafine-grained materials, prepared via severe plastic deformation of bulk solids, had high strength but limited ductility when tested at low temperatures. This could be remedied by severe plastic deformation to very high strains, followed by immediate short-term annealing. The appearance of acceptable ductility was then attributed to grain-boundary sliding even at relatively low temperatures; something which was possible because of the presence of a large number of non-equilibrium grain boundaries and an associated increased diffusivity.

In another review of the topic[5] it was pointed out that some alloys are actually weakened by severe plastic deformation processes such as equal-channel angular pressing or high-pressure torsion: spray-cast AA7034 alloy can have its grain size reduced from 2.1 to 0.3μm by 6 angular-pressing passes but can still exhibit an overall weakening due to partial loss of a hardening phase. Weakening can also occur in Zn-22%Al eutectoid alloy when precipitates of the stable hexagonal close-packed zinc phase, which is contained within aluminium-rich grains in the annealed condition, are absorbed by the zinc-rich grains during processing. Similar weakening is found in Pb-62%Sn and Bi-42%Sn eutectic alloys. On the other hand, there can be an additional strengthening effect because the processed materials may contain nanoparticles, nanotwins, non-equilibrium grain boundaries and grain-boundary segregation; all of which may impart higher strengths than expected on the basis of the usual Hall-Petch relationship. In the case of nickel, for example, electrodeposited material has grains which have no dislocation sub-structure and exhibits normal Hall–Petch behaviour. In processed nickel, a higher strength is observed which can be directly traced to contributions arising from low-angle grain boundaries, high-angle grain boundaries and non-equilibrium grain boundaries. In processed metals, excess dislocations are present in the grain boundaries and impart extra energy to the grain boundaries, thus rendering them non-equilibrium. The contribution of these dislocations to the strength, and their interaction with segregants, leads to strengths which are above the Hall-Petch values and impart so-called super-strength. AA1570, AA7475, Ti-6Al-4V and 0.4%C steel also exhibit super-strength. The super-strength phenomenon led to the realisation that there are 2 critical lengths in the Hall–Petch relationship. In ultrafine-grained material, the Hall–Petch exponent can vary from -1/2 to -1 in some regions while, in another region, the exponent can be -3. This gives a positive slope and imparts exceptional hardening. In other cases, there can be a softening at very small grain sizes. The strengthening is attributed to the presence of appreciable grain-boundary segregation at grain sizes below about 100nm.

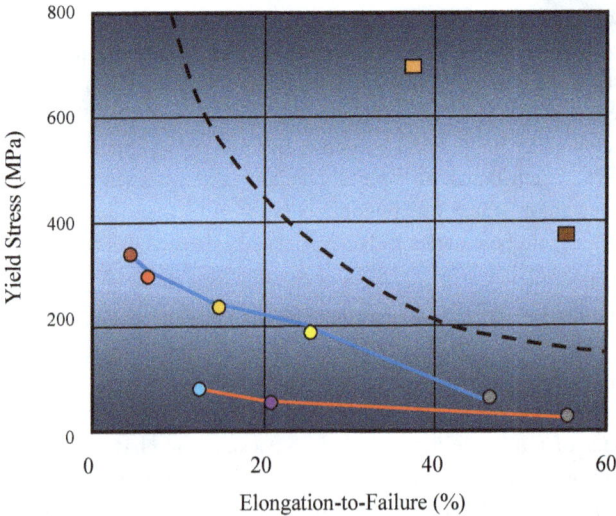

Figure 1. Most ultrafine-grained metals are confined below the dashed line, but the titanium and copper points escape because the samples were prepared using severe plastic deformation to very large strains. Blue line: copper, red line: aluminium. Dark red point: 61% rolling strain, blue: 75%, light red: 37%, purple: 35%, orange: 21%, yellow: 11%, gray: 0%. Light brown square: nano-titanium, dark-brown square: nano-copper.

Copper

The alternative term, paradox of severe plastic deformation, was applied[6] to the occurrence of pre-existing deformation twins in ultrafine-grained copper which had been produced by equal-channel angular pressing and heat treatment. The high strength of the material was attributed to a high density of coherent twin boundaries which acted as obstacles to gliding dislocations. The twins favoured increases in the dislocation density, both in the grains with twins and in grains without them. The sample therefore hardened, further augmenting its strength. At the same time, it exhibited a high ductility, in that the deformation behavior was controlled mainly by the boundaries of grains which were free from twins.

Accumulative roll bonding was used[7] to produce ultrafine-grained sheet which was then annealed (300C, 1h) in order to optimize its strength and ductility. Homogeneous lamellar ultra-fine grained material with a thickness of 200 to 300nm was found following 6 passes of accumulative roll bonding. The microhardness and tensile strength of as-treated copper increased (table 1), while the ductility and strain-hardening decreased with cumulative deformation-strain. The as-treated specimens fractured in a macroscopically brittle and microscopically ductile manner. Following annealing, discontinuous recrystallization occurred in the neighbouring high strain-energy interface; before that in the matrix. The recrystallization-rate increased with increasing cumulative strain. The annealed 6-pass treated material had a completely recrystallized microstructure, with grain sizes ranging from 5 to 10μm. The annealing reduced the microhardness and tensile strength, but improved the ductility and strain-hardening of the ultrafine-grained copper. The as-annealed material fractured in a ductile manner, with predominant dimples and shear-zones. The presented results were compared (figure 2) with those resulting from the use of other methods of deformation, such as dynamic plastic deformation, quasi-static deformation, cryorolling and equal-channel angular pressing; followed by annealing. The presented data points were located in the same region as previously published data. The figure shows that the uniform elongation for coarse-grained copper without casting artefacts is about 51%. Conventional strengthening mechanisms such as grain refinement and deformation increase the yield strength at the expense of ductility. Nanostructured and ultrafine-grained copper, prepared by using dynamic plastic deformation and equal-channel angular pressing, as well as cryorolling, had a 5 to 10 times higher yield strength than that of coarse-grained copper while the uniform elongation was less than 5%. Annealing increased the uniform elongation, but with a decrease in yield strength. This common paradox arose from plastic deformation which was dominated by dislocation-slip. Grain-refinement and deformation increased the yield strength by increasing the critical shear stress for slip-initiation. Ductility was

Materials Research Forum LLC
https://doi.org/10.21741/9781644903230

related not only to dislocation nucleation, but more closely related to slip kinetics. Grain boundaries and deformation-induced dislocation cells blocked dislocation slip and multiplication and thus reduced ductility. Dislocation-controlled plastic deformation governed the existence of the strength–plasticity paradox. Following a single pass, multiple slip-systems were activated simultaneously and a large number of dislocations accumulated in the form of entangled structures containing a high density of dislocations, and elongated sub-structures comprising dislocation cells and sub-grains having low-angle grain boundaries. The cell-structures were such that most dislocations were entangled in the cell wall, leaving a relatively low density of dislocations within the cell interior. The elongated cellular structure made up of individual dislocation-tangles was not parallel to the rolling direction, but was at an angle of some 40° to it. The average cell thickness ranged from 300 to 800nm. The cell boundaries were mainly low-angle grain boundaries. Following 3 passes, a large number of geometrically necessary dislocations appeared in the deformed microstructure in order to satisfy strain-compatibility. Lamellar ultra-fine grains formed along the rolling direction and the fraction of high-angle grain boundaries increased. Following 6 passes of cumulative rolling, the ultra-fine lamellar grains – with a thickness of about 200nm - acquired sharp grain boundaries; with a low dislocation-density within the grains, due to dislocation-recovery and dynamic recrystallization. Only small numbers of deformation-twins were observed, suggesting that grain-refinement was dominated mainly by dislocation-slip segmentation during accumulative rolling deformation. Deformation-twinning occurred when the grains reached the ultra-fine regime. Following the annealing (300, 1h) of 1-pass rolled material, the microstructure contained a small fraction of recrystallized grains and a large fraction of deformed sub-structures. Partial recrystallization over an area of some 25% occurred in high-strain regions, and abnormal grain growth was observed. In annealed 3-pass material, the area of recrystallization was about 62.5%, and the size of the recrystallized grains ranged from 5 to 15μm. In annealed 6-pass material, newly recrystallized grains had completely replaced the deformed microstructure and their size, of 5 to 10μm, was smaller than that (50μm) of the original material. Numerous annealing twins formed in the recrystallized grains. Discontinuous recrystallization could occur in regions of high dislocation-density and cube texture. Cube-oriented grains exhibited a higher growth rate than that of other orientations and the formation of a cube orientation thus resulted in discontinuous recrystallization and abnormal grain growth in the case of 1-pass material. When compared with copper which was subjected to equal-channel angular pressing, the recrystallized grains in the roll-processed copper were not uniform, due to the inhomogeneously deformed microstructure. The heterogeneously deformed microstructure was attributed to the unequal matrix deformation, due to the uneven distribution of shear forces acting along the thickness direction during the rolling process.

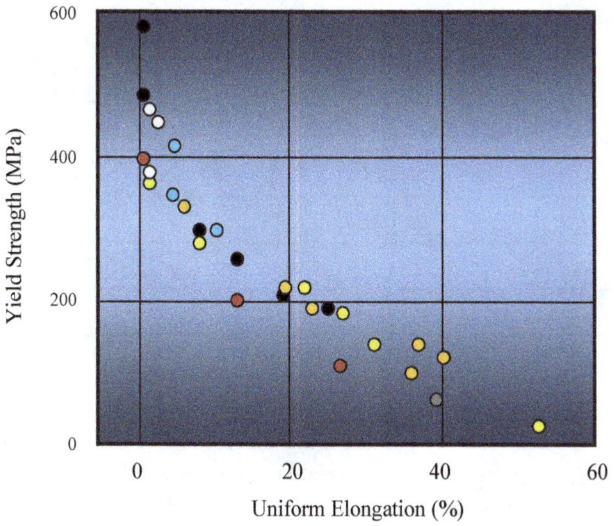

Figure 2. Yield strength versus uniform elongation for copper prepared using various methods and subsequent annealing. Black: dynamically plastically deformed copper, gray: cryorolled copper, white: present accumulative roll bonding, blue: present accumulative roll bonding plus annealing, orange: equal-channel angular pressed copper, yellow: equal-channel angular pressed copper plus annealing, brown: quasi-statically deformed copper.

Table 1. Yield strength, ultimate tensile strength, uniform elongation and elongation-to-failure of accumulative roll-bonded and as-annealed copper samples

Material	YS(MPa)	UTS(MPa)	e_u(%)	e_f(%)
ARB, 1 pass	380	390	1.9	5.4
ARB 3 passes	450	468	2.0	4.3
ARB 6 passes	460	482	1.8	4.5
Annealed ARB 1 pass	350	363	4.0	9.2
Annealed ARB 3 passes	410	423	4.7	9.0
Annealed ARB 6 passes	300	337	10.3	16.5
Untreated, coarse-grained	70	220	40	42.4

Hydrostatic tube cyclic expansion extrusion was used[8] for the severe plastic deformation-forming of ultrafine-grained tubes by pushing a mandrel through a hollow-tube sample and applying high fluid pressures. Commercial-purity copper tube was processed, showing that the grain size was reduced to below 150nm by a single pass, starting from an initial value of 65μm. The yield and ultimate tensile strengths were increased to 270MPa and 345MPa, respectively by one pass, starting with initial values of 75MPa and 207MPa, respectively. The elongation was reduced from 55% to 41%. The microhardness was also increased, from 59HV to 133HV.

Pure copper compacts with a bimodal, so-called harmonic structure, were prepared[9] by mechanical milling, followed by spark plasma sintering at 873K for 1h under 100MPa. In such an harmonic structure, a coarse-grained core is embedded within a matrix of 3-dimensional and a contiguous connected shell of ultrafine-grained material, The subsequent thermomechanical processing involved 50% reduction by cold-rolling and annealing (673K, 0.5h). This processing greatly increased the strength of both this harmonic-structured copper, and that of homogeneous copper, because recrystallization refined the grain size. The ductility was also increased to some extent. The harmonic-structured and thermomechanically processed compact exhibited (table 2) a yield strength of 125MPa and an ultimate tensile strength of 251.2MPa; increases of 32% and 24% with respect to the equivalent properties of the homogeneous copper. The strain-hardening ability of the former material was higher than that of the latter material. The coarse-grained core provided high ductility, by storing dislocations, while the fine-grained shell directly increased the strength and indirectly improved the ductility because high stresses within the contiguous shell delayed the point of plastic instability. Increasing the shell fraction was therefore an effective approach to obtaining materials possessing both high strength and high ductility. Following thermomechanical treatment, the harmonic-structured copper exhibited considerable increases in yield strength (>125MPa) and ultimate tensile strength (>251.1MPa), as well as in uniform elongation (>37.1%) and fracture elongation (>43.6%), as compared with non-processed samples. The thermomechanical processing greatly contributed to strength and ductility, due mainly to recrystallization which refined the grain size and eliminated any internal stresses acquired during sintering and severe plastic deformation. Strain-hardening is required as a stabilizing mechanism to control shear-banding instability and delay or spread necking. The harmonic-structured copper exhibited a higher strain-hardening than did homogeneous copper. The processing also improved the strain-hardening ability, as reflected by the better mechanical properties of processed material: coarse-grained regions were constrained by skeleton-like shell regions where high stress-localization existed, leading to a higher yield strength. The complex deformation behaviour in the core region led to strengthening of the harmonic structure and delayed plastic instability, as compared with the homogeneous material. The homogeneous copper had an average grain size of about 42.5μm, and thermomechanical processing produced an average grain size of 20.1μm. The harmonic structure was of micron-scale but was homogeneous at the macro-scale, consisting of ultra-fine grains and fine grains smaller than 5μm. Coarse grains, larger than 5μm, made up the core. The thermomechanical processing minimized the grain size during cold rolling, especially at the shell/core boundary, leading to an increase in the shell fraction. The harmonic-structured copper exhibited more fine grains, especially ones smaller than 5μm, than did the homogeneous copper.

Table 2. Tensile properties of processed homogeneous copper, harmonic-structured copper and processed harmonic-structured copper

Material	YS(MPa)	UTS(MPa)	e_u(%)	e_f(%)
processed homogeneous	95	202.5	43.9	63.0
harmonic-structured	110	236.1	32.0	39.0
processed harmonic-structured	125	251.2	37.1	43.6

Oxygen-free copper of better than 3N5 purity was subjected[10] to equal-channel angular pressing at room temperature, for up to 24 passes, and pulled to failure at room temperature using strain-rates ranging from 10^{-4} to 10^{-2}/s. Severe plastic deformation processing methods, such as equal-channel angular pressing, are effective in reducing the grain size and imparting high strengths, but ultra-fine grain metals tend to have very limited ductilities or elongations-to-failure. It is nevertheless possible to obtain both high strength and high ductility if the material is subjected to very high strains. Metals which are processed by using conventional techniques such as rolling, drawing or extrusion exhibit low ductility or low yield strength while metals with nanocrystalline structures, produced by severe plastic deformation, possess good strength and good ductility. As previously not, this phenomenon was first noted for 4N6 copper which was subjected to 16 passes of equal-channel angular pressing at room temperature, and for 3N8 titanium subjected to 5 revolutions of high-pressure torsion at room temperature. Both processing methods produced grain sizes of about 100nm. Tensile testing at a strain-rate of 10^{-3}/s measured elongations-to-failure of about 51% and 43% for copper and titanium, respectively. The degree of micro-strain in the present case increased with strain following lower numbers of passes, but decreased following between 16 and 24 passes. Similar trends were exhibited by the dislocation density, the Vickers microhardness and the tensile yield stress. There was a minor increase in the crystallite size at the high strains which were imparted by 24 passes, and dynamic recovery occurred at the highest strains. Tensile testing at a strain-rate of 10^{-3}/s resulted in a yield stress of about 391MPa and an elongation-to-failure of 52%. The generally low ductility in materials with small grain sizes was due to a low strain-rate sensitivity and a low rate of strain-hardening. The latter was limited because small grains offer only limited dislocation-storage. Dislocations therefore move easily through the grains and are absorbed in grain boundaries. Elongations are thereby limited for ultra-fine grain structures, even if the metal is ductile when it is coarse-grained.

In a similar vein, severely plastically deformed pure copper microstructures were produced[11] by subjecting cylindrical samples to high-pressure torsion. A crystal-plasticity plus continuum dislocation dynamics model was used to predict the mechanical behaviour of the samples, and took account of mechanisms which were based upon gradients of dislocation density and grain size, of the back-stress fields of grain boundaries, of dislocation-density transmission across grain boundaries and of stress/strain gradient effects.

A theoretical study was made[12] of the way in which precipitates can both inhibit dislocation glide and lead to hardening and low ductility, but also act as efficient dislocation sources under high stresses and thus promote ductility. A model was based upon the observation that dislocation nucleation by a nanoprecipitate is controlled by the lattice-mismatch between the precipitate and matrix. A combination of the Orowan mechanism for precipitate hardening, and a critical condition for dislocation nucleation at the nanoprecipitate, was expected to indicate the optimum precipitate size and its spacing. Molecular dynamics simulations were used to investigate the nucleation of dislocations by nanoprecipitates in copper, with the embedded-atom potential being used to describe the atomic interaction between a gold precipitate and the lattice of a perfect copper single crystal. The size of the periodic simulation-box was 530nm, 187nm and 0.8nm for 2-dimensional simulation and 66nm, 51nm and 25nm for 3-dimensional simulations. Noting that a perfect dislocation in a face-centred cubic crystal glides on a (111) plane, along its <110> Burgers vector, the x, y and z coordinates of the simulation box were oriented parallel to the $[11\bar{2}]$, $[111]$ and $[1\bar{1}0]$ crystallographic axes, respectively. By replacing a few of the copper atoms with gold atoms, model nanoprecipitates were created in the face-centred cubic structure. Following energy-minimization and structural relaxation, a shear strain of ε_{yz} was applied at a constant rate of 2 x 10^9/s. When dislocations began to be nucleated, the applied strain was kept constant. Initial temperatures of 1K or 300K were assumed in order to determine the effect of heat. The atomic structure changed when dislocation nucleation occurred, and these changes were tracked by using a common-neighbour analysis of the nucleated atoms. The critical nucleation state could be characterised by counting the number of distorted atoms. The simulations showed that nanoprecipitates are an ideal dislocation-source under high stresses and are very different in action to conventional Frank-Read sources in crystalline materials.

Rotary swaging was used to prepare copper wire having a fibre texture and with long ultra-fine grains lying along the wire axis[13]. Annealed 3N8-purity copper rod was subjected, at room temperature, to a high hydrostatic stress and a strain-rate of about 1/s. A series of dies was used to decrease the diameter of the specimen, with true deformation strains of 0.5, 1.0, 1.5, 2.0 and 2.5. During swaging, the original 54μm equiaxed coarse grains were elongated, in the axial direction, into super-long columnar grains with an average diameter of 2.06μm and a length of 339μm at a true deformation strain of 2.5. The swaging also introduced strong <111> and weak <100> fibre textures along the copper axis, together with a high density of dislocations: 9.19 x 10^{14}/m^2 at a true strain of 2.5. Some of the dislocations self-organized into dislocation-cell boundaries and formed sub-structures or sub-grains, with low-angle grain boundaries within the elongated coarse grains. The dislocation cells had an average width of 220nm and a length of 25.1μm. With increasing true strain, the intensities of the <111> and <100> textures, the dislocation density and the volume fraction of low-angle grain boundaries first greatly increased and then saturated, or decreased slightly when the true strain was 2.5. Uniaxial tensile tests were performed at a strain-rate of 10^{-3}/s at ambient and liquid nitrogen temperatures. The coarse-grained copper had a yield strength of 60MPa and a tensile elongation-to-failure of 57%. The swaging gradually increased the yield strength to 450MPa and decreased the ductility to 10% at a true strain of 2.5. In swaged material, necking occurred immediately following yielding, due to an almost complete absence of strain-hardening ability. The swaging deformation absorbed the strain-hardening by saturating dislocation accumulation. The ductility was enhanced to 13%, and the strain-hardening and uniform elongation were improved, when tested at liquid nitrogen temperatures. Annealing (573K, 2h) increased the ductility to 20% while maintaining a yield strength of 380MPa. The improved ductility and uniform elongation were attributed to the enhanced dislocation accumulation capability. The main strengthening mechanism involved the high density of dislocations.

Finite-element crystal plasticity simulations, taking account of size-dependent dislocation storage, grain-boundary sliding and void-induced damage in a Gurson-type model were used[14] to clarify the fundamental mechanism underlying the strength-ductility compromise in nanocrystalline metals. The finite-element model for non-textured polycrystals assumed a randomly-distributed equiaxed structure with a grain-boundary thickness of 1nm and a volume-weighted average grain diameter of 14.5nm. It was found that the tensile failure strain of cube-textured copper having a given grain size could be more than double that of the failure strain for non-textured copper in the presence of high grain-boundary strengthening. Molecular dynamics simulations had demonstrated that the fracture of nanocrystalline face-centred cubic metals was governed by the nucleation, growth and coalescence of voids. The macroscopic yield strength of nanocrystalline metals, deduced using the Gurson model, agreed well with those simulations over a wide range of void-sizes. The use of grain-boundary strengthening, texture manipulation and the creation of gradient microstructures was a possible strategy for optimising both strength and ductility.

Magnesium

Surface milling using a form of ball-peening was applied[15] to 3N-purity magnesium, showing that high-frequency and multi-directional deformation strains were thereby introduced and created gradients in grain-size and orientation as a function of depth from the surface. There was also a gradient in the density of twin meshes (intersecting arrays of twins) along the sample thickness. Room-temperature tensile tests showed that, following milling, the material had a higher ultimate tensile strength, together with a doubled ductility. It was shown that the twin meshes could be responsible for the activation of additional slip systems and a greater strain-hardening, leading to a higher uniform plastic strain.

Nickel

In order to try to circumvent the strength-ductility compromise, a strategy was essayed[16] which involved controlling the density and distribution of dislocations within the as-prepared metal. An electroplating technique was used to insert just 2vol% of nanoscale (7nm) domains that then spread out within nickel grains. The pulsed electrodeposition procedure was such that the disturbance caused by high current-density pulses, together with a supply of grain-refining agent, initiated numerous nanoscale domains which were crystallographically misoriented by a few degrees with respect to the matrix grain. The domains did not become contiguous because their growth was interrupted when the plating current was suddenly reduced to a level which grew ultrafine grains. The nanoscale domains consequently appeared intermittently within larger grains. The nanodomain nickel was then as strong as a nanocrystalline material, and simultaneously as ductile as a coarse-grained metal. This permitted the material to 'escape the envelope' (figure 3). The marked strengthening was attributed to the blocking of nanoscale domains, quite low volume-fractions of which acted as precipitates and obstructed gliding dislocations. This then increased the strength to that of nanocrystalline nickel. These deliberately dispersed domains, with a large population of dislocations, simultaneously promoted multiplication mechanisms while leaving room within the grains for dislocations to entangle and be stored. This led to a marked strain-hardening rate at flow stresses of the order of gigapascals, and supported uniform elongation to the levels typical of face-centred cubic metals. The exceptional strength x ductility values made the nanodomain nickel far better than previous materials.

Aqueous sulfamate-based electrolyte was used to produce nanocrystalline nickel by using direct-current electrodeposition[17]. The nickel, with a thickness of 180 to 200μm, was initially deposited onto a stainless-steel substrate. Room-temperature tensile tests were performed using strain-rates ranging from 0.00005 to 0.01/s. Isolated large grains, with a size of 120nm, were surrounded by grains which were smaller than 30nm. This inhomogeneous microstructure was similar to those resulting from abnormal growth during low-temperature annealing. Dislocations, twins and sub-grain boundaries were present within the large grains. The nanocrystalline nickel had an average grain size of 27.2nm, with the distribution ranging from 5 to 120nm. The presence of $MnCl_2$ in the electrolyte led to a broad grain-size distribution in the deposits. The average 0.2%-offset yield strength was 638MPa, the average ultimate tensile strength was 1176MPa and the average elongation-to-fracture was 10.6%. In comparison with nanocrystalline nickel with a grain-size distribution of 10 to 50nm and average grain size of 15 to 25nm, with an ultimate tensile strength of 1200 to 1300MPa and a ductility of 4 to 5%, the present samples had a greatly increased ductility and similar strengths. The average grain size following deformation increased by 19.3nm and there was a sharp decrease in the number-fraction of sizes below 30nm. A few grain coalescences larger than 200nm were observed in post-deformation microstructures. It was suggested that the large grains acted so as to merge neighbouring small grains via grain rotation during plastic deformation. Dislocations which were trapped during grain coalescence, and dislocation pile-ups at grain boundaries were observed. This indicated that intracrystalline dislocation-sliding predominated. Dislocations could barely move in small nanograins because very high stresses were required to activate Frank-Read sources and, for grain sizes of 80 to 120nm, these were estimated to be 498 to 746MPa. Frank-Read sources could be activated in grains larger than 80nm. Grain growth during plastic deformation could favour the activation of Frank-Read sources with increasing deformation strain.

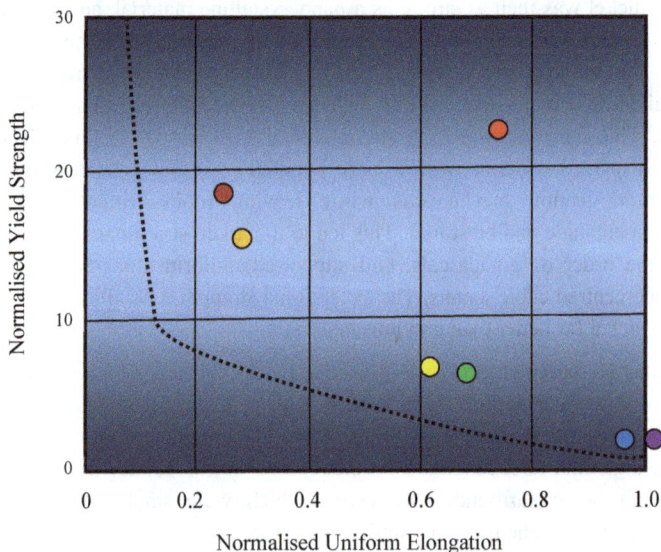

Figure 3. Normalized yield strength versus normalized uniform tensile elongation for various nanostructured metals (below dotted line). Brown: nanotwinned copper, orange: consolidated copper, yellow: gradient steel, green: bimodal copper, blue: gradient copper, purple: gradient transformation-induced plasticity steel. The nanodomain nickel exhibits nanocrystal-level strength plus coarse-grain level ductility.

Titanium

A study was made[18] of the crystallographic textures and deformation mechanisms which characterise the ω-phase of titanium when subjected to high-pressure (6GPa) torsion at 298K. The specimens were discs with a radius of 20mm and a thickness of 1.5mm in the as-received condition, or after 0.1, 0.5, 1 and 5 turns under torsion. The original material had an average grain-size of 10μm. A zone which was located 5mm away from the centre of the sample (half the radius of the disc) was studied. Seven slip and twinning systems were chosen, including: basal {00•1}<11•0>, prismatic {10•0}<12•0> and first-order pyramidal {10•1}<12•0> with a Burgers vector of 1/3<11•0>, first-order pyramidal {10•1}<11•3>, second-order pyramidal {112}<11•3> with a Burgers vector of c + a = 1/3<11•3>, tensile {10•2}<10•1> and compression {21•2}<21•3>. Taking account of the possible activation of all of the possible slip families and twinning systems, each slip or twinning system was accorded a relative value of critical resolved shear stress. The choice of active slip and twinning systems was based upon calculation of the minimum work expended on deformation. As a result of varying the relative values of critical resolved shear stress, the barrier to some slip/twinning systems was lessened while others became more difficult. The lattice parameters of ω were such that c/a = 0.608. In order to identify the activation mechanism of a slip or twinning systems, one type of slip/twinning system was assigned a least relative critical resolved shear stress of 1, while a 10-fold higher value was assigned to the other systems. As a result, the other systems took part in the deformation, but their contribution to texture development was minimal. As demonstrated previously, during the high-pressure torsion of ω-Ti, beginning at 0.1 of a turn, a phase-change occurred; the volume fraction of which increased with the number of turns. Quantitative phase analysis showed that, if the fraction of ω-phase was about 1% after 0.1 of a turn, it had increased to about 33% after 0.5 of a turn and to about 58% after 5 turns. After this, the ω-phase became the main one in severely plastically deformed titanium. During the increasing severity of the deformation, basal slip and tensile twinning were the active processes. The results showed that the ω-phase of titanium deformed in the same manner as α-Ti. It was concluded that the most active slip-systems were the pyramidal <a + c> ones of first and second order. The activity of the pyramidal <a> slip systems was noticeable only in the initial stage of severe plastic deformation. High-pressure torsion caused activation of the twinning processes by tension, although their contribution to deformation was minimal.

Bulk nanostructured titanium was produced[19] by means of equal-channel angular pressing and constant or alternating high-pressure torsion. A simulation model for the flow stress was developed which took account of deformation via dislocation slip and by twinning. Twinning appeared to play an important role in the deformation of nanostructured materials when below a critical grain size. The flow stress was assumed to be the sum of stresses associated with dislocation slip and with twinning. An additional hardening due to twinning was incorporated via a Hall-Petch dependence. Actual deformation of titanium was studied in coarse-grained material with an average grain size of 30μm and in nanostructured material produced by 10 equal-channel angular pressing passes at 723K at a rate of 6mm/min, followed by 5 turns of high-pressure torsion. The model adequately explained the hardening behaviour, and differences in the hardening characteristics of various nanostructures could be attributed to differences in the dislocation densities in cell-interiors and in cell-walls, and to varying concentrations of deformation-induced vacancies. Further attempts were made[20] to apply a dislocation–kinetic approach to the analysis of strength and ductility in metals and alloys in general. The influence of temperature and strain-rate upon the character of dislocation processes was investigated. Special attention was paid to the effect of grain boundaries, twins, impurity atoms and any non-equilibrium state of the grain boundaries upon the activation of deformation mechanisms in bulk ultrafine-grained metallic materials produced via severe plastic deformation.

A previously unidentified heterogeneous-lamella structure was found[21] to exhibit the strength of an ultrafine-grained material and the ductility of a coarse-grained material. The structure was produced by asymmetrical rolling, followed by partial recrystallization. The rolling elongated the originally equiaxed grains into the lamella structure; which was heterogeneous, with some areas having a finer lamella spacing than others. This was caused by a variation in the slip system and plastic strain in grains of differing initial orientation. The rolling produced a higher strain near to the sample, and a consequent nanostructured surface layer. There was a slight structural gradient, before and after recrystallization. During partial recrystallization, lamellae having a finer structure recrystallized so as to form soft microcrystalline lamellae. Other lamellae recovered, while retaining a hard ultrafine-grained structure. Most of the recrystallized grains clustered into long lamellae lying along the rolling direction. The recrystallised grains were equiaxed and almost dislocation-free, while the smaller ultrafine-grained grains still contained a high density of dislocations. The volume fraction of recrystallised grains decreased with depth, but the grain size slightly increased. The slight gradients in microstructure, before and after recrystallization, led to microhardness variations. Heterogeneous lamellae titanium, with a thickness of 60μm and 80μm was as strong as ultrafine-grained titanium and as ductile as coarse-grained titanium, respectively. Both of the heterogeneous lamellae structures were 3 times as strong as coarse-grained titanium, while retaining the same uniform tensile elongation. All of the other samples, regardless of thickness were also much stronger and more ductile than was coarse-grained titanium (figure 4). The ultrafine-grained titanium became mechanically unstable soon after yielding. It was concluded that the heterogeneous lamellae material derived their high ductility from a high strain-hardening rate, which became much higher than that of coarse-grained titanium following plastic straining. The heterogeneous lamellae structure was regarded as being a sort of bimodal structure; one which was much more effective in producing strain-hardening than was a conventional bimodal structure. The elongated inclusions produced a higher strain-hardening than did spherical ones, particularly when their long axis was aligned with the loading direction. The lamellar form made the mutual constraint between soft and hard lamellae more effective, leading to higher back-stresses. The full constraint of the soft lamellae by the hard lamellae matrix made it more effective, in constraining plastic deformation of the soft lamellae and in developing higher back-stresses, than was a conventional bimodal structure. The heterogeneous lamellae structure had a high density of interlamellar interfaces at which dislocations could pile up and increase back-stress and dislocation hardening. Back-stresses were the main cause of the high strengths. Back-stress hardening and dislocation hardening were responsible for the high strain-hardening rate and resultant high ductility.

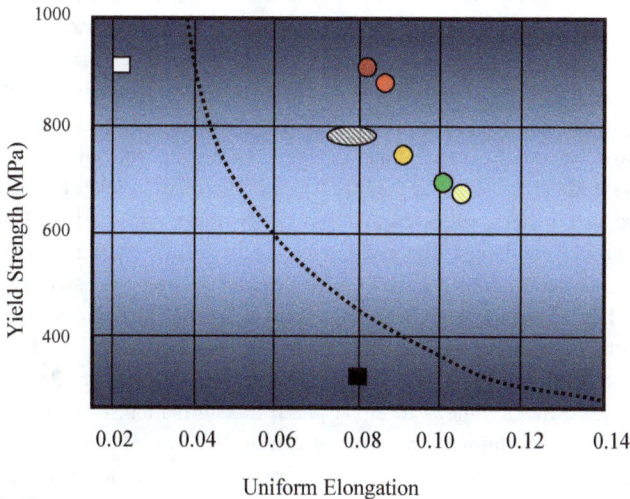

Figure 4. Yield strength versus uniform tensile elongation of heterogeneous-lamellae titanium. Conventional titanium alloys lie below the dotted line. Black square: coarse-grained titanium. White square: ultrafine-grained titanium. Circles are, brown: 60µ heterogeneous lamellae, red: 80µm, orange: 100µm, green: 200µm, yellow: 300µm heterogeneous lamellae. Cross-hatched oval: Ti-6Al-4V.

Commercial-purity 2N3 titanium was subjected[22] to high-pressure torsion, revealing that in the case of cold-consolidated powder its mechanical behaviour depended upon phase inhomogeneity (the presence of alpha and omega phases), structural inhomogeneity (a bimodal grain size distribution) and spatial inhomogeneity (the presence of retained porosity). A high strength of high-pressure torsion bulk titanium was due to the formation of the hard omega phase during processing at room temperature. That phase transformed back into nanograined alpha phase domains during short annealing times at high temperatures. The high-pressure torsion consolidation of titanium powder led to the formation of brittle specimens which exhibited high strengths but essentially zero plasticity. Disks of bulk titanium were processed at room temperature, using an applied pressure of 6.0GPa, a rotation-speed of 1rpm and torsional straining through 1 or 5 turns. Titanium pre-compacts were densified by applying a pressure of 6GPa for 60s. Dense disks were further processed in the same manner as bulk titanium. The bulk and densified powder, following annealing (700C, 1h, vacuum), had mean grain sizes of 142.8µm and 85.1µm, respectively. The microhardness of bulk titanium typically increased by 4 to 5 times, with increasing numbers of revolutions. Following pure 6.0GPa compression, without torsional straining, and annealing (700C, 2.4ks) the powder exhibited a very inhomogeneous microhardness distribution across the disk diameter. Following high pressure torsion processing through 5 whole revolutions, the microhardness was almost homogeneous and equal to about 300HV; close to that of processed bulk titanium. Following high-pressure torsion consolidation, the crystallite size and micro-strain were comparable to those of processed bulk titanium. Following annealing (250 or 300C, 600s), the crystallite size increased slightly and the micro-strain level fell to 0.001. The microstructure of powder, processed at room temperature, consisted of 100nm grains with blurred grain boundaries. The grain size of processed bulk titanium was slightly smaller than that of consolidated powder. Samples which were processed using 5 whole revolutions exhibited an increase in strength and a significant decrease in elongation. These were attributed to the alpha-omega phase transformation, and to the consequent presence of the brittle nanocrystalline omega phase. When testing at 250C, above the temperature (175C) of the omega to alpha phase transformation, there was no unusual mechanical behavior. Additional annealing at 250 or 300C for 600s led to a decrease in strength but a significant increase in ductility, from less than 0.1 to about 0.3. This was explained in terms of an omega-phase transformation back to nanocrystalline (~10nm) alpha phase, thus promoting high strength. In the case of processed consolidated titanium powder, all of the specimens, except the annealed (300C, 600s) ones, had zero ductility. Pre-compacted samples (6.0GPa, zero torsion) which were annealed (700C, 2.4ks) had an elongation of about 0.006 and a Young's modulus of 88.9GPa. These values were lower than those for coarse-grained alpha titanium, and this was attributed to the retained porous structure of the compressed powder. Consolidated powder, processed by torsional straining to 5 whole revolutions, had a Young's modulus of about 136.8GPa; higher than that of coarse-grained material. The increased Young's modulus of nanostructured materials was related to the high internal stresses which were introduced by the processing. Annealing at 250C brought the Young's modulus back to the level (117.2) of coarse-grained titanium. This material exhibited a plasticity of some 2.2%; lower than the 8% of consolidated powder following 10 whole revolutions.

Commercially pure 2N7 titanium was subjected[23], upon recrystallization (700C, 2h) to 5-step hydrostatic extrusion and 2-stage rotary swaging at room temperature. The mechanical properties and microstructure were investigated following heat treatment at 200C. Although the material was annealed, deformation-induced defects were still present within the microstructure thus proving that continuous dynamic recrystallization predominated in the later stages of processing. Regardless of the test conditions, the material lacked homogeneity, with a range of grain sizes and shapes and the presence of various deformation-induced defects. A considerable increase in the ductility of a heat-treated material compensates for the decrease in tensile properties. Low-temperature annealing may thus provide some initial solution to the strength-ductility problem in materials which are subjected to severe plastic deformation. The latter process is based upon the principle of work-hardening, in that the mechanical properties are increased due to the accumulation of microstructural defects such as point defects, grain boundaries and dislocations. Heat treatment might easily restore ductility, but that tends to lead to a loss of strength.

The tensile strength and ductility of commercially-pure titanium in various conditions were determined[24], the conditions being untreated or annealed (700C or 900C, 1h). The as-received material (table 3) had an ultimate tensile strength of 344MPa, a yield strength of 375MPa, an elongation of 20% and a modulus of elasticity of 105MPa. Greater oxidation occurred at 900C and this temperature, being above the beta transus temperature, led to the formation of acicular martensite. There was a decrease in the tensile strength when annealing at 700C, but it recovered when annealing at 900C. The 700C-annealed material exhibited the highest ductility (50%). Because 700C was below the transformation to beta-phase, that material was slightly more ductile than the 900C-annealed material, but the latter had a lower tensile strength. The 900C-annealing was above the transformation temperature (883C) to the beta phase. The ductility increased and the tensile strength returned to the original value of 376MPa.

Table 3. Strength and ductility of commercially-pure titanium in various conditions

Condition	UTS(MPa)	Elongation(%)	Reduction-of-Area(%)
as-received	375	41	75
annealed (700C, 1h)	360	50	82
annealed (900C, 1h)	376	47	65

Titanium laminates possessing good strength-ductility combinations were prepared[25] by means of the temperature-controlled rolling and annealing of as-sintered multilayered pure titanium foils with grain-size gradients. A gradient-structure, with the average grain size decreasing from the centre layer, with an alternating stacking of equiaxed fine-grain bands and elongated-grain bands, to the surface layer, with equiaxed fine grains, was first formed by optimizing the rolling process. A generalized Schmidt-factor analysis indicated that slip-dominated and twin-dominated grain-refinement mechanisms operated during rolling. As well as prismatic slip, two types of contraction-twinning could be activated so as to permit grain-refinement in domains where slip systems were difficult to activate. Following annealing, equiaxed fine grains in the laminates did not coarsen appreciably, while the elongated grains in the centre layer partially merged and grew to become elongated coarse grains. The microstructure of the centre layer thus had a typical heterogeneous lamellar structure consisting of an alternating stacking of elongated coarse grains and fine-grain bands. Together with the grain-size gradient, a multi-scale heterostructured titanium laminate was formed. The as-rolled laminate had an ultimate tensile strength of 930MPa and an elongation-to-fracture of 12.3%. The annealed (425C) laminate had an UTS of 823MPa and an elongation-to fracture 22.4%. It was suggested that annealing markedly reduced the dislocation density in the elongated coarse grains of the laminates, so that more geometrically necessary dislocations could accumulate during testing; leading to hetero-deformation induced strengthening plus an additional hardening due to the unique laminate structure. This evaded the strength-ductility paradox.

Attention was focused on the stresses which drive local plastic deformation in layered materials such as titanium/titanium, titanium/aluminium and aluminium/aluminium[26]. The latter were prepared by alternately stacking foils of AA1060, commercially-pure aluminium, and commercially-pure titanium and subjecting them to hot-pressing, hot-rolling and annealing. There were marked differences between the constituent layers. The grain size of coarse-grained layers was some 5 times larger than that of fine-grained layers. In comparison with homogeneous analogues, good mechanical properties were offered by the layered metals. The yield strength of titanium/titanium layers exceeded the predictions of the Hall-Petch and rule-of-mixtures laws. The fracture strain of titanium/aluminium layered materials was better than that of either constituent layer. The aluminium/aluminium layers offered an enhanced combination of yield strength and uniform elongation when compared with homogeneous aluminium. Ahead of the yielding of titanium/titanium layers (212MPa), there was an early local yielding due to dislocation activity near to the layer interface at an applied stress of 160MPa. It was proposed that the layered structure introduced local stress concentrations at its interfaces and promoted additional deformation mechanisms in titanium. There was also intrinsic local stress-partitioning during deformation due to the incompatibility of the constituent layers. The deformation of titanium/aluminium layers before the applied stress reached 175MPa could be classified into 2 stages. During the first stage, both the titanium and the aluminium deformed elastically but the differing elastic moduli led to an unequal local stress-partitioning between titanium and aluminium via Hooke's law. The aluminium then yielded before the titanium due to the lower yield strength, and any further increase in applied stress was carried mainly by the titanium rather than the aluminium. The mismatch in Poisson ratios also generated local stresses perpendicular to the loading direction. The existence of a multi-axial local stress state was anticipated in layered materials, even when uniaxial stresses were applied, and this could activate deformation mechanisms that deviated from predictions only of the applied stress. Local strains were pivotal in regulating deformation mechanisms.

Tungsten

Nanocrystalline tungsten having a strength of some 3.0GPa (quasi-static compression) and about 4.0GPa (dynamic compression) was prepared[27] by means of high-pressure torsion at 500C. The dynamic uniaxial compression was performed by using a Kolsky bar or split Hopkinson pressure bar system which imparted strain-rates of 0.0001 to 0.001/s. The grain boundaries were mainly of high-angle type, and were atomically sharp, non-equilibrium and high-energy. Edge dislocations were present within the grains and these, combined with low impurity concentrations along pre-existing grain boundaries, were suggested to increase the ductility. Under dynamic compression, the material underwent localized shearing, followed by cracking and failure. This behaviour was similar to that exhibited by ultrafine-grained material which had been subjected to equal-channel angular pressing and cold-rolling. It was also similar to the behaviour exhibited by other body-centred cubic metals having ultrafine-grained or nanocrystalline microstructures. The shear-band width in the high-pressure torsion tungsten was much narrower (<5μm) than that (circa 40μm) in ultrafine-grained material.

Non-Ferrous Alloys

Aluminium

Layers of AA3003 and AA1060, or of AA1060 alone, were studied[28] via the *in situ* monitoring of local strain variations. Large microstructural and textural variations could lead to a large difference in the transverse compressive strain between the layers when under uniaxial tension, suggesting that appreciable deformation incompatibility and greater strain-gradients were caused by constraint effects. One of the samples consisted of layers of non-annealed and annealed aluminium foils (AA1060) having a thickness of 200μm. The non-annealed AA1060 foil had a typical rolling texture, and an average grain size of 2.7μm. Annealing (500C, 1h) produced complete recrystallization. The non-annealed and annealed foils were stacked alternately and the laminate was hot-pressed (400C, 45MPa, 1h). The composite sheet was then rolled at 300C to a thickness reduction of about 65%, and annealed (250C, 1h). Other layered samples were prepared by the 200C roller-bonding of 1.5mm AA3003 and AA1060 sheets to a thickness reduction of 50%, and annealed (610C or 550C). Control samples were prepared from non-annealed AA1060 foils alone in the same way, from annealed AA3003 sheets alone and from annealed AA1060 sheets alone. They comprised fine-grained and coarse-grained layers, with a similar layer thickness (75μm). There were differing microstructures and textures in the constituent layers. The fine-grained and coarse-grained layers of the AA3003/AA1060 composites had average grain sizes of 1.1 and 9.3μm, respectively. The corresponding grain sizes in the AA1060/AA1060 composites were 4.2 and 13.0μm, respectively. The fine-grained layers of both types of composite had strong S {123}<634> and brass {110}<112> textures. The coarse-grained layers of the AA3003/AA1060 composites had a shear ({100}<011> texture with a weak {100}<001> cube texture. The recrystallized microstructures of the layers contained a high proportion of high-angle grain boundaries. There were many other subtle differences, but the important factor was that the layered aluminium composites offered better strength–ductility combinations. The strengths of the 2-layered materials helpfully obeyed the law-of-mixtures. The AA3003/AA1060, with its greater variability of microstructure and texture of the constituent layers, had a higher work-hardening rate than that of AA1060/AA1060 before the strain reached ~0.125. The strain-hardening rate of AA3003/AA1060 then fell rapidly to ~100MPa. The more constant strain-hardening rate of AA1060/AA1060 meant that it offered a better resistance to plastic instability and thus a higher elongation-to-fracture than that of AA3003/AA1060; being better than that of coarse-grained bulk AA1060. The layered structures optimised the strength–ductility compromise by increasing the work-hardening rate and the resistance to plasticity instability. Large numbers of interfacial microcracks were found in both layered materials. Intra-layer microcracks nucleated mainly at the interfaces between the aluminium matrix and Al-Mn eutectic phases. The intra-layer could aid the transverse propagation of interfacial microcracks across the fine-grained layers of AA3003/AA1060, leading to the formation of primary cracks.

Samples of $Al_{81}Cu_{13}Si_6$, $Al_{88}Cu_8Si_4$ and $Al_{92}Cu_{5.6}Si_{2.4}$ were first prepared by gravity-casting[29]. Rod-shaped samples were then directionally solidified at 4.0mm/s in a temperature gradient of 17K/mm. The $Al_{81}Cu_{13}Si_6$ had a microscale binary eutectic embedded within an ultrafine ternary eutectic matrix, and a yield strength of up to 750MPa. It failed via brittle fracture. The $Al_{92}Cu_{5.6}Si_{2.4}$ had a hypereutectic composite structure and exhibited both a high tensile strength and a high ductility. Room-temperature tensile tests were performed at an initial strain-rate of 5 x 10^{-4}/s. All 3 alloys comprised α-Al solid solution, θ-Al_2Cu intermetallic and a primary silicon phase. The $Al_{81}Cu_{13}Si_6$ comprised a few micrometre-scale dendrites, and lamellar cellular structures were randomly distributed in nanoscale eutectic structures. The cellular structures consisted of a binary eutectic of α-Al and θ-Al_2Cu. The ultrafine ternary eutectic matrix consisted of α-Al, θ-Al_2Cu and β-Si. There was an obviously hypoeutectic structure in the case of $Al_{88}Cu_8Si_4$. The high (52%) volume fraction of primary phase consisted of α-Al solid solution, with the grain size ranging from 5 to 10μm. Some ultrafine ternary eutectic was also found. With increasing aluminium content, the volume fraction and grain size of the α-Al phase increased to 65% and 20 to 30μm, respectively, with only a little ultrafine ternary eutectic remaining. The $Al_{88}Cu_8Si_4$ exhibited an obvious ductility, and a tensile strength of 600MPa. The plasticity was much greater for $Al_{92}Cu_{5.6}Si_{2.4}$, which exhibited a strain of 10% and a tensile strength of 500MPa. The α-Al played a pivotal role in determining the mechanical properties of these alloys. With increasing volume fraction of the α-Al, the ductility generally increased. A larger grain size of α-Al also tended to prevent the propagation of microcracks or shear bands; leading to an improvement in ductility. The work-hardening ability of the soft primary phase was also expected to have a marked effect upon the plasticity of these alloys. It was suggested that their optimum mechanical properties could be obtained by tailoring the volume-fraction, morphology and work-hardening ability of the α-Al phase. The $Al_{81}Cu_{13}Si_6$ exhibited cleavage-like features under tensile stress, with just one main crack running through the whole specimen and little plastic deformation. A rough fracture surface in the case of $Al_{88}Cu_8Si_4$ appeared to reflect a different fracture behaviour, in which ductile deformation occurred before fracture. The primary dendritic α-Al played an important role in improving plasticity. As the volume fraction and grain size of dendritic α-Al increased, more dimpled patterns were observed and were continuously distributed in the $Al_{92}Cu_{5.6}Si_{2.4}$.

Materials Research Forum LLC
https://doi.org/10.21741/9781644903230

Experiments were carried out[30] on cast Al-7wt%Si hypoeutectic alloy which was annealed (445K, 5h, vacuum) before high-pressure torsion deformation. The annealing imparted an average initial α-aluminium grain size of 135μm, with silicon particles having an average size of 2.8μm. The high-pressure torsion was performed under quasi-constrained conditions at 298 or 445K, using a uniaxial compressive pressure of 6.0GPa through 1/4, 1, 5 or 10 turns of the anvil-like holders. The average equivalent tensile strains, for 0, 1/4, 1, 5 and 10 turns were 0, 2.3, 9.2, 46.2 and 92.5, respectively. Specimens were tested under tension at 298K, using strain-rates of 0.0001, 0.001, 0.0033 or 0.01/s. The strain due to grain boundary sliding was estimated, and it was found that the processing had produced very marked grain-refinement, even after one quarter of a turn. After 10 turns, the grain size was 400nm. Less refinement occurred during processing at 445K, but the final grain size was 1.5μm after 10 turns (table 4). The fraction of high-angle grain boundaries (those with misorientations greater than 15°) was determined, revealing a sudden decrease at the onset of processing at 298 or 445K. This was attributed to the introduction of numerous new sub-grain boundaries with low misorientation angles following torsion through one quarter of a turn. As the number of turns increased the fraction generally increased, apart from a slight decrease, at room temperature, between 5 and 10 turns. That was due to the cyclic occurrence of sub-grain formation, grain refinement and recrystallization. The strength increased with processing at 298K, while the ductility initially decreased after one quarter of a turn, in comparison to that of the as-cast material. After 5 turns the material exhibited both high strength and high ductility. Good, but lesser, strength and ductility were observed after 10 turns at 445K. A low fraction of high-angle grain boundaries imparted a lower ductility, while a high fraction led to high ductility. At a strain-rate of 0.001/s, the minimum elongation-to-failure was 36% and the maximum elongation was 75%. The contributions which grain boundary sliding made to the total strain was calculated (table 5). It was concluded that imparting both high strength and high ductility required both a very small grain size and a very high fraction of high-angle grain boundaries. This latter was required because high ductility was a result of the occurrence of localized grain boundary. Such sliding in turn occurred only when the grain boundaries had high misorientation angles.

Materials Research Forum LLC
https://doi.org/10.21741/9781644903230

Table 4. Average grain size of tensile samples of high-pressure torsion processed Al-7%Si

Processing Temperature (K)	Turns	Average Grain Size (μm)
298	0.25	5.3
298	1	1.1
298	5	0.5
298	10	0.4
445	0.25	122
445	1	16
445	5	1.8
445	10	1.5

Table 5. Contribution of grain boundary sliding to the total strain, normalized UTS and normalized elongation-to-failure of high-pressure torsion processed Al-7%Si

Turns	Strain-Rate (/s)	e_t(%)	UTS/UTS$^{as\text{-}cast}$	$e_t/e_t^{as\text{-}cast}$
0.25	0.001	5	2.0	0.9
1	0.001	13	2.3	1.2
5	0.001	14	2.4	1.9
10	0.001	17	2.0	2.1
1	0.01	4	1.9	0.7
10	0.01	14	2.1	1.2

The underlying cause of the strength-ductility paradox was explored[31] from the point-of-view of interactions between the thermodynamics and kinetics of phase transitions, and the deformation involved in materials processing, by using age-hardenable aluminium alloys as examples. Hardening and softening effects were analysed in terms of energy accumulation and dissipation during precipitate nucleation and growth, together with dislocation multiplication and annihilation. The key thermokinetics of driving forces and energy barriers were deduced from atomistic calculations. Analytical models for linking mechanical performance to thermokinetics in terms of microstructural characteristics were derived from phase transition and deformation theories. This had implications for precipitate control, as seen in a 6XXX alloy at the prescribed aging temperature. In general, the yield strength could be increased by increasing the critical shear stress for dislocation-glide initiation. This then weakened the strain-hardening capability and reduced the ductility. The strength could be described by, $\sigma = \alpha\mu bM\rho^{0.5}$, where ρ was the total dislocation density, while the plastic strain, ε, was given by $\varepsilon = b/M^{-1}$, where α was a material constant, μ was the shear modulus, b was the Burgers vector, M was the Taylor factor and l was the dislocation slip distance. An increased σ was associated with an increased ρ; required to overcome the resistance arising from dislocation interactions. The thermokinetics of phase transition and deformation were seen to be synergistic, rather than being independent of one another. At the atomic scale, phase transition proceeded via the diffusion of atoms or shearing. Deformation was triggered by twinning or dislocation glide within the lattice. via a displacement of about one atomic spacing. A generalized stability concept was proposed on the basis of a correlation between thermodynamics and kinetics which acted as a decreased driving force, associated with an increased energy barrier, and *vice versa*. The compromise between strength and ductility was therefore analogous to the synergy between the thermodynamics and kinetics of materials processing.

The paradox was investigated[32,33] in the case of cast Al-7%Si alloy which had been subjected to high-pressure torsion, using up to 10 turns at 298 or 445K. This process reduced the grain size to a minimum of about 0.4µm, and decreased the average size of the silicon particles. Samples which were given high numbers of high-pressure torsion turns revealed both a high strength and a high ductility when tested at relatively low strain-rates. The strain-rate sensitivity under these conditions was about 0.14, thus implying that flow occurred via limited grain-boundary sliding and crystallographic slip. There was thus suggested to exist the possibility of obtaining strength and high ductility by increasing the number of high-pressure torsion turns.

Hypoeutectic AlSi11Cu samples were prepared[34] via laser powder-bed fusion and subjected to severe plastic deformation. The initial material had a structure consisting of an aluminium matrix, surrounded by a silicon-enriched cellular network. The yield strength was 393MPa, the ultimate tensile strength was 565MPa, the uniform elongation was 4.6% and the elongation-to-fracture was 4.9%. Following deformation using equal channel angular pressing, the elongation had almost doubled (12%), the yield strength was 319MPa and the ultimate tensile strength was 494MPa. The lower mechanical properties of the processed material were attributed to the breakdown of the interconnected network. The breakdown created space for dislocation motion, and new particles had a pinning effect which supported a higher tensile ductility than that of the initial material. The average grain size had been reduced from 10μm to 1μm by 6 angular-pressing passes. The microstructure had changed from one composed of columnar grains to a heterogeneous microstructure with ultra-fine (200nm to 500nm) plus elongated grains (5μm to 10μm). Superplastic behaviour was favoured by strain-rates of about 0.001/s and a temperature of 400C. Elongations of more than 70% were found for the severely deformed material, but it did not exhibit superplastic behaviour following the 6 passes. This was attributed to the heterogeneous microstructure, which was not entirely ultrafine-grained.

In situ mechanical testing within a transmission electron microscope revealed[35] that nanoscale helium bubbles could simultaneously increase the strength and ductility of small-volume monocrystalline Al-4Cu pillars and evade the strength-ductility paradox. The nanoscale bubbles acted as internal dislocation sources and shearable obstacles, thus promoting dislocation-nucleation and storage, and leading to a higher flow stress and plasticity plus a greater uniform deformation of the Al-4Cu pillars. The nanoscale bubbles were also quite stable under plastic strain, and their coarsening and coalescence were observed only in the final localized deformation (necking) stage.

Bars of commercial AA1050 and AA7075 were subjected[36] to equal channel angular pressing, after annealing (480C, 5h) to create an homogeneous solid solution and quenching in room-temperature water. An effective strain of about 1 was imposed per pressing-pass. Uniaxial tensile tests were performed at room temperature, using an initial quasi-static strain-rate of 0.001/s. Processed ultrafine-grained AA7075 exhibited randomly oriented equiaxed grains with an average size of about 800nm. Spherical or cylindrical precipitates, with diameters of tens of nanometres and interparticle distances of tens to hundreds of nanometres, were observed in the grain interiors but rarely at grain boundaries. The precipitates were mainly those of the non-coherent stable η-phase. The processing altered the η-phase orientation and its interfacial energy with respect to the aluminium matrix, such that the initially cylindrical η-phase evolved into equiaxed particles. The dislocation density in most of the grains was low, and some of the dislocations were pinned by the precipitates. Low-angle grain boundaries, formed from polygonised dislocation walls, were frequently observed. It was recalled that ultrafine-grained material, processed at high temperatures, contained a higher fraction of low-angle grain boundaries than did material which was processed at room temperature. This was attributed to the kinetic recovery of dislocations into dislocation walls at higher temperatures, and was consistent with the low observed dislocation density and the fact

Materials Research Forum LLC
https://doi.org/10.21741/9781644903230

that pinned dislocations alone remained following deformation. The ultrafine-grained AA1050 exhibited randomly oriented equiaxed grains with an average size of 700nm. Most of the grains were free from dislocations, and were greatly misoriented with respect to surrounding grains; suggesting that they formed via dynamic recrystallization during processing at room temperature. Some grains contained entangled dislocations, dislocation forests and discrete single dislocations. The ultrafine-grained AA7075 had a yield strength of 350MPa, an ultimate tensile strength of 500MPa, a uniform elongation of 18% and a tensile ductility of 19%. Ultrafine-grained AA1050 had a yield strength of 170MPa, an ultimate tensile strength of 180MPa, a uniform elongation of 2.5% and a tensile ductility of 7%. In this material, necking occurred immediately after yielding and was attributed to a lack of strain-hardening and dislocation storage after the severe plastic deformation. The presence of nanoscale precipitates in the ultrafine-grained AA7075 increased the yield strength and the tensile ductility by promoting strain-hardening. The latter was enabled by the impeding effect of nanoscale precipitates upon dislocations. The fracture surfaces of ultrafine-grained AA7075 were irregular, with a high concentration of uneven concavities and protrusions. The material fractured in a ductile manner, as indicated by an homogeneous distribution of dimples and a fracture reduction-of-area of 35%. The dimple-size of 100 to 500nm was comparable to the interparticle spacing of η-phase precipitates. Spherical particles were found at the base of the dimples. Ultrafine-grained AA1050 also fractured in a ductile manner, with a reduction-of-area of 40% and numerous dimples. It was noted that a high strain-hardening rate and a high strain-rate sensitivity were required to impart a high tensile ductility because they prolonged elongation by delaying necking instability. The strain-rate sensitivity was in turn related to the flow-stress activation volume and the thermally-activated slip mechanism. Strain-hardening was due mainly to interactions with other dislocations and with lattice defects. Reducing the grain size into the ultrafine range minimised the space available for dislocation accumulation and multiplication, and almost completely removed the strain-hardening ability. The introduction of a high density of nanoscale precipitates, as in AA7075, restored the strain-hardening ability by augmenting the dislocation-accumulation ability. The introduction of nanoscale precipitates into the ultrafine-grained AA7075 matrix reduced the tendency to shear fracture and rendered the deformation more uniform, with a higher tensile ductility.

Copper

A multi-scale approach was used[37] to study the severe plastic deformation of bulk copper ingots which were subjected to equal-channel angular pressing. The macro-level took account of material flow and crystallographic texture. The meso-level considered the size and shape of grains, their size-distribution and their mutual misorientations. The micro-level covered structural defects and their interaction. It was found that the radii of curvature of the channels, and the friction coefficient, affected the strain-rates and the preferred crystallographic orientations imposed on bulk copper ingots. In later work it was noted[38] that some models of severe plastic deformation considered the appearance of stages IV and V of strain-hardening. The Estrin-Tóth model involves stage IV and the concept of a 2-phase composite structure with cell-walls having a high dislocation density and cell interiors with a relatively low dislocation density; all within a cubic cell structure. An Estrin-Tóth dislocation model was used to describe the high-pressure torsion of copper. Changes in dislocation-density were studied in cell boundaries and in their interiors. The concentrations of vacancies produced during deformation, the annihilation of dislocations during non-conservative motion and the generation of dislocations during multiple cross-slip were all considered. The effect of hydrostatic pressure was analysed with regard to limitations on lattice diffusion. With increasing strain, the collective behaviour of dislocations was such that a fragmented cell structure appeared which was characterized by high angles of fragment misorientation. Fragmentation occurred against the background of a so-called frozen cell structure. It was noted that the activation energy (79.2kJ/mol) for grain-boundary diffusion in ultrafine-grained copper was much lower than the value (107kJ/mol) which was typical of diffusion along stationary grain-boundaries in coarse-grained copper. An absence of dislocation accumulation could also be attributed to the activation of recovery processes which partially involved dislocation cross-slip. In further work[39], the mechanisms governing the strength and ductility of nanostructured copper and its alloys when subjected to high-pressure torsion were analysed in terms of an elastic-plastic model. The effects of the Peierls strength, of solid-solution hardening, of dislocation-hardening and of twinning-hardening, while also taking account of annihilation processes, were estimated. In the case of Cu-5at%Al, annihilation processes contributed to deformation, with the material being hardened by the accumulation of dislocations at twin boundaries and thus achieving a higher ultimate tensile strength. In the case of Cu-16at%Al, the annihilation processes were limited, thus deformation was limited and the degree of homogeneous deformation was lower in comparison with that of Cu-5at%Al. A much higher concentration of deformation-induced vacancies also contributed to failure of the former alloy.

An improved Estrin-Tóth model was in fact used[40]. The structure of a severely plastically deformed material was assumed to consist of cell-blocks which were separated by boundaries that contained excess dislocations that were 'fluffed' by non-excess sessile dislocations. Such a structure was termed fragmented. Dislocations within the fragment interior formed a cell structure, with the total density of immobilized dislocations in the cell-walls and cell-interiors being equal to ρ_c. By taking account of the total dislocation density, ρ_w, and the density of non-excess sessile dislocations, ρ_f, in the fragment boundaries it was possible to calculate the density of excess dislocations within the boundaries, and the misorientations between the fragments. The annihilation of dislocations during non-conservative motion controlled by vacancy diffusion and the annihilation of edge dislocations was taken to explain the absence of hardening in severe plastic strain. Material flow was supposed to involve dislocation generation during multiple cross-slip. Tensile tests of annealed (550C, 1h) pure copper in the as-received state, and of copper following 1, 4 and 8 passes of equal-channel angular pressing, were performed at room temperature using an initial strain-rate of 0.00055/s. The plastic flow of samples was heterogeneous along the sample length, leading to localization of the plastic strain. The elongation-to-failure of the as-received material and of samples following 1, 4 and 8 passes was 45%, 15%, 10% and 11%, the yield stress was 227MPa, 376MPa, 436MPa and 460MPa, and the failure stress was 415MPa, 600MPa, 680MPa and 740MPa, respectively. For the purpose of calculation, a polycrystalline sample was assumed to consist of 830 weighted orientations, and that up to 12 octahedral slip systems, {111}<110>, typical of a face-centred cubic lattice were operating The threshold shear stress was the same for all of the slip systems. Simulations were performed using a true strain of 0.51, where the cross-section of the neck decreased by some 1.6 times. It was concluded that the model consistently described the experimental data, and permitted a quantitative description of the dependence of the dislocation density upon the number of equal-channel angular pressing passes. It was shown that the role of annihilation processes, controlled by the capture and non-conservative motion of edge dislocations, was reduced following the 8th pass; the mechanism of dislocation generation being suppressed by the forest dislocations. These changes reflected a qualitative reconstruction of fragment boundaries, with the latter becoming essentially impermeable and the fraction of non-excess sessile dislocations becoming negligibly small. Structure refinement occurred mainly up to the 4th pass, after which the misorientations between fragments remained low. The misorientations between fragments began to increase following the 4th pass. A marked increase in vacancy concentration occurred with increasing number of passes, and the dislocation density in fragment boundaries as well as the total dislocation density increased. The model predicted the changes in vacancy concentration in the fragment boundaries and permitted the estimation of their role in strain-hardening. The insignificant strain-hardening of materials subjected to equal-channel angular pressing was attributed to the fact that the increase in source activity was compensated by an increase in annihilation processes. A considerable increase in the yield stress following 1, 4 and 8 passes was attributed to an increase in the dislocation density.

Further kinetic modelling[41], performed using the dislocation model, considered all of the possible deformation mechanisms and evaluated the processes affecting the grain boundaries in copper subjected to severe plastic deformation. It was again demonstrated that the processes occurring in the grain boundaries were sensitive to the applied hydrostatic pressure. Temperature changes affected the activation or suppression of deformation processes. These processes were more intensive in ultrafine-grained states. In some cases, temperature increases and hydrostatic pressure increases promoted the action of Frank-Read sources. The efficiency of grain boundaries as sinks for dislocations increased. The annihilation of screw dislocations during double cross-slip, and their annihilation during non-conservative motion both increased. In other cases, the same effects were observed during hydrostatic pressure increase and temperature decrease. In particular, the concentration of deformation-produced vacancies markedly increased. The same factors favourably affected structure refinement and the increase in misorientations between grains. The last factor was related to a reduction in the fraction of non-redundant sessile dislocations in the boundaries. The fraction immobilized in boundaries thus increased and they became almost impenetrable. The quantity which was most sensitive to temperature and hydrostatic pressure changes was the dislocation density in the grain boundaries of the ultrafine-grained copper. Both a temperature reduction and a hydrostatic pressure increase led to its growth and to an increase in flow stress.

Bulk metallic glass samples of $Cu_{46}Zr_{47}Al_7$ in plate form were prepared[42] by copper-mould suction-casting and subjected to surface mechanical attrition treatment by using a 20kHz system in which a form of shot-peening bombarded the material so that surface plastic deformation might alter the local structure and residual stresses. The shot was 10μm to 300μm in diameter and, because of the high frequency of the system, the entire surface of the material was hit a very larger number of times within a short period. The shot velocities were of the order of 10m/s and could impose strain-rates of 1000/s on the metallic glass. Following the treatment, the surface was roughened, with a root-mean-square roughness of 360nm. Profuse sub-surface shear-banding was observed (table 6). The size of the treatment-affected zone increased with treatment-time, but saturated at 210mm after 24ks. The overall structural amorphousness was retained even after 1h of the surface mechanical attrition treatment. Tensile tests were performed at room temperature, using a strain-rate of 0.0001/s. The as-cast material did not exhibit any tensile ductility, and fractured instantly upon yielding. The treated samples exhibited various degrees of unusual strain-hardening and thus some tensile ductility before final failure. A tensile ductility of some 2% was found when the material had been treated for

15ks. The ductility decreased however when the bulk metallic glass was under-treated or over-treated. The yield strength of treated material decreased slightly, but the fracture strength markedly increased, with increasing tensile ductility. Regardless of the tensile ductility, all of the samples finally fractured in the same way; that is, cracking occurred along an inclined plane. Hardly any shear-banding was visible on the surface of as-cast samples, while the outer surfaces of treated samples were covered with a large number of shear bands. This suggested that the tensile ductility of the treated glass was probably the result of multiple shear-banding. The tensile ductility could then be explained in terms of an effective shear-offset mechanism in which the tensile plastic strain was assumed to be accommodated by shear-band displacements that formed on the sample surface (table 7). The simply-calculated plastic strain in fact largely agreed with the measured plastic strain, thus confirming that multiple shear-banding was the main source of tensile ductility. The surface mechanical attrition treatment naturally produced residual stresses in the treated material. By using the hole-drilling method, the residual stress distribution in the sample-thickness direction was measured for as-cast and treated material. The most ductile glass, following 15ks treatment, exhibited a very different residual stress distribution and peaked at a compressive stress of about 500MPa at 30mm from the impacted surface. It then changed quickly to the tensile residual stress. The other residual stress profiles were either rather flat, with the maximum compressive residual stress ranging from 250 to 2100MPa, or exhibited a broad peak which was located much deeper than was the 15ks peak. The residual stress profiles of shot-peened crystalline metals tended to saturate with dislocation density and treatment-time. The marked variation in residual stress profiles, as found in treated metallic glass, was therefore unusual and suggested that structural evolution was triggered by the treatment. Severe plastic deformation, driven by the surface mechanical attrition treatment, could lead to so-called rejuvenation or further disordering of the glass structure, together with volume dilatation. The outer layer of the treated glass would then expand relative to the inner layer and generate compressive residual stresses near to the surface while leaving tensile residual stresses within. It was proposed that structural relaxation could occur with increasing treatment time, and counteract the rejuvenation, thus homogenizing the disorder of the amorphous structure and smoothing the residual stress profile, as was observed. The enthalpy of crystallization of the treated material was essentially the same as that of the as-cast material, confirming the overall amorphous nature of the treated material. The relaxation enthalpy increased markedly during treatment, from 23.54J/g for as-cast material to 25.86J/g for material treated for 15ks. The relaxation enthalpy decreased, and tended to saturate at about 25.2J/g as the treatment was continued. Transmission electron microscopy revealed a grain-like microstructure of dark/bright contrast at about 10mm below the impacted surface. The size of dark regions ranged from 100 to 200nm while those of bright ones ranged from 10 to 50nm. At a depth of some 30mm, the size of dark regions increased to 2 to 10mm while the size of bright regions remained almost unchanged. The gradation of microstructural features ceased at 120mm. All of the microstructures were amorphous. Nano-indentation tests showed that the hardness decreased as the indentations approached the impacted surface. This supported the suggestion that the amorphous microstructure formed due to profuse shear-banding. It was concluded that brittle fracture in bulk metallic glasses under tensile loading could be overcome by introducing a graded amorphous structure using the surface attrition process. Due to the high volume fraction of liquid-like atoms near to the surface, shear-banding was promoted while cavitation and fracture were suppressed. This ensured a tensile ductility which was not possible in the as-cast material. The tensile-ductility increase was associated moreover with an increased fracture strength, thus evading the strength-ductility paradox.

Table 6. Average shear-band spacing, shear-band density, shear-band offset and shear-angle in surface mechanical attrition treated $Cu_{46}Zr_{47}Al_7$ bulk metallic glass

Treatment Time(ks)	Spacing(μm)	Density(/μm)	Offset(μm)	Angle(°)
0.6	34	0.029	0.2	55
15	11	0.091	0.2	52-55
24	16	0.061	0.2	54-56
36	28	0.036	0.2	53-55

Table 7. Measured and calculated plastic strains of surface mechanical attrition treated $Cu_{46}Zr_{47}Al_7$ bulk metallic glass

Treatment Time (ks)	$e_p^{calculated}$(%)	$e_p^{measured}$(%)
0.6	0.34	0.81
15	1.07	1.5
24	0.68	0.89
36	0.41	0.79

Hot-rolled Cu-0.4Cr-0.3wt%Zr alloy strips were subjected[43] to various thermo-mechanical processes and the results revealed that a desirable combination of strength and ductility could be obtained by optimising the sequence of cold rolling and aging. Treatment which involved aging followed by cold rolling was preferable to cold rolling with aging. An ultimate tensile strength of 568MPa was obtained by aging (450C. 3h), followed by 80% cold rolling at room temperature. A high ductility was attributed to the twin lamellar structure.

Copper-aluminium-nickel material (Cu-5Al-3wt%Ni) with ultra-fine grains that contained precipitates was prepared[44] from 3N-purity elements. The stacking-fault energy was less than $7erg/cm^2$. Following homogenization (900C, 2h), it was subjected to unidirectional cold-rolling at room temperature, Split-Hopkinson pressure-bar deformation and aging. Following aging, the strength and elongation had increased to 797.06MPa and 28.66%. With increasing aging temperature, the tensile strength first increased and then decreased. The tensile strength attained the above maximum value when the aging temperature was 350C. As usual, a high strain-hardening rate was critical for a good uniform elongation as it helped to delay local deformation. The optimizing factors for this alloy were twinning, grain boundaries and second-phase particles. The cooperation of these mechanisms promoted the improvement of strength and ductility. Large numbers of dislocations and twin structures were introduced during Split-Hopkinson pressure-bar deformation, due to the high strain-rate of 2500/s. Phase-field methods were used to model dynamically the interaction between twins and dislocations. The typical characteristics of high strength-ductility copper alloys were a small grain-size, a high twin-density and stable grain boundaries which offered more space for the unified improvement of strength and ductility. The Zener-Hollomon parameter characterizes the creation of twins. Increasing the strain-rate and decreasing the deformation temperature could increase its value, and suppress dynamic recovery via dislocation cross-slip and climb. This led to a higher twin-density and a finer grain-size. The Split-Hopkinson pressure-bar can impose a higher strain-hardening rate than can other severe plastic deformation techniques. This deformation-method must however be combined with post-annealing because fine grains and an initial high density of dislocations leave little room for further dislocation accumulation, resulting in reduced ductility.

Magnesium

Equal-channel angular pressing was applied[45] to AZ31 magnesium alloy with added CaO at high temperatures, including 1 pass at 523K, 2 passes at 523K and 4 passes with a combination of the first 2 passes at 523K and the following 2 passes at 473K. A few steps of reduction led to a refinement in grain size to about 1.5μm after 6 passes. The equivalent grain size of the CaO-free alloy was 2.2μm. Room-temperature compression tests showed that the equal-channel angular pressed CaO-containing material exhibited an increased yield strength which was higher than that of as-processed commercial AZ31 alloy. Both materials retained their ductility, in spite of a marked reduction in grain size. The improved strength of the AZ31-CaO alloy was attributed to the formation of fine Al_2Ca precipitates which broke up during the equal-channel angular pressing and accelerated microstructural refinement. The retained ductility was attributed to an increased strain-hardening capability of the AZ31 alloy at room temperature.

When equal-channel angular pressing at 593K was applied to extruded AZ80[46], the original grain size of 20μm was reduced by 75%, by the first pass, to about 5μm. The $Mg_{17}Al_{12}$ phase which was originally distributed along the grain boundaries as networks, was rearranged into a more homogeneous distribution, and acquired a more spherical shape. No strengthening was observed, but the grain refinement and precipitate-fragmentation resulted in a 28% increase in elongation-to-failure during tensile testing. There was no further change in microstructure during 6 passes. The size of the precipitates increased slightly during processing. The yield stress of the as-extruded alloy was greatest, with a 10% decrease occurring following the first pass. It remained at the same level during further processing. The ultimate tensile strength exhibited a similar trend. The elongation-to-failure instead increased from about 14 to 18% following the first pass. The as-extruded material had a texture in which (002) was parallel to the extrusion direction. The texture gradually became randomized during processing, and this was attributed to rotation of the sample during the equal-channel angular pressing. Some favourably oriented grains, those having higher resolved shear stresses on their basal planes, could undergo plastic deformation at a lower applied stress; resulting in a lower yield stress than that for as-extruded material with its (002) planes parallel to the tensile axis. Rapid water-cooling immediately following pressing could produce a supersaturated solid solution in the matrix and precipitates, leading to increased lattice parameters. The lattice spacing of the magnesium matrix was increased by 0.05 to 0.1%. The as-extruded material exhibited a typical brittle cleavage fracture in which the grain size was clearly visible. Following processing, the fracture surfaces exhibited very small dimples which were commensurate in size to the grain size. This implied that grain-refinement might be the cause of the ductility improvement.

The combination of a high damping capacity and a high strength is a loosely-related paradox which arises in magnesium alloys: a high damping capacity requires the easy movement of dislocations on the basal plane, while a high strength requires that dislocations be impeded by second-phase particles, grain boundaries or stacking faults.

With regard to damping capacity, a large grain-size benefits the movement of dislocations, and manganese can hinder the growth of grain boundaries. The addition of manganese to magnesium, together with a low (165 to 300C) extrusion temperature, here[47] created material with a bimodal grain structure that comprised fine recrystallized grains and coarse non-recrystallized grains. The fine grains encouraged the activation of grain-boundary slip, thus imparting good plasticity. The large numbers of parallel dislocations in the non-recrystallized grains, due to dislocation recovery and the dynamic recrystallization rate, were suppressed by the low extrusion temperature. Manganese precipitates meanwhile impeded the movement of dislocations, leading to a high work-hardening rate. This not only led to a high yield strength, but also improved the room-temperature damping capacity. Because of the non-recrystallized grains, the relevant internal-friction peak increased, and moved to a lower temperature. When the extrusion temperature was 220C, the overall properties of a Mg-1Mn alloy were superior to those of high damping-capacity Mg-0.6Zr alloy. The yield strength, plasticity and damping capacity at $\varepsilon = 0.001$ were 169MPa, 32.5% and 0.042.

As-cast Mg-2Y-0.6Nd-0.5Zr alloy was forged at 350C and subjected to various numbers of passes of equal-channel angular pressing[48]. The grain size was thereby reduced from 140μm to 1.9μm and acquired a uniform fine microstructure due to dynamic recrystallization during the processing. The forged material had a pronounced basal texture while the equal-channel angular pressed material exhibited a tilted texture, with the basal orientation being weaker and the cylindrical and pyramidal orientations being stronger. The alloy possessed a combination of tensile strength (262MPa) and elongation (22.4%) following forging and 4 passes of equal-channel angular pressing. These values were 118% and 113% better, respectively, than those of the as-cast material. The strength decreased slightly due to textural changes and excessive recrystallization during the angular pressing. The plasticity of samples which were subjected to 6 passes was increased to 25.2%, while the maximum pole density was reduced to 7.02 from the forged-state value of 13.87. Due to casting defects and large grain sizes, the ultimate tensile strength and elongation of the as-cast alloy were relatively low (table 8). Forging markedly increased the UTS to 244MPa while the elongation was increased to 12.3%. Grain refinement improved the strength and plasticity, but work-hardening decreased the plasticity, so that the overall plasticity of the forged alloy increased less. The strength unexpectedly decreased with decreasing grain size, contrary to the Hall-Petch effect. This was attributed to the fact that the mechanical properties of magnesium alloys following equal-channel angular pressing were dependent not only upon the grain size but also upon the texture. Following one pass, the basal texture was greatly reduced and this reduced the strength of the alloy. The softening effect due to dynamic recovery and textural changes outweighed the strengthening effect due to grain-size reduction. The ductility improved with grain refinement, and the strength and plasticity of the alloy were greatly improved by the angular pressing. Following 4 passes, the tensile strength and elongation of the alloy continued to increase. The grain size after 6 passes was very fine, and exhibited a marked ability to coordinate and deform with reduced dislocation participation. The high degree of recrystallization reduced the dislocation density and led to the strength being lower than that following 4 passes, while the plasticity increased further.

Table 8. Grain size and properties of Mg-2Y-0.6Nd-0.5Zr

Sample	Grain Size (μm)	UTS (MPa)	Elongation (%)
as-cast	140.00	120.00	10.50
as-forged	25.00	244.00	12.30
1-pass	14.80	233.00	17.40
2-passes	4.50	247.00	19.70
4-passes	3.00	262.00	22.40
6-passes	1.90	258.00	25.20

Nickel

The yield strength of nickel-based superalloys is particularly interesting in the present context because it tends to increase with temperature to up to 800C and, as usual, the increase in yield strength is generally associated with a decrease in ductility. The main feature of the microstructure is a dispersion of precipitates, having an ordered $L1_2$ structure, that is coherently embedded in a solid-solution strengthened γ-phase matrix. Due to presence of the ordered precipitates, nickel-based superalloys exhibit the above yield-strength anomaly. The substructural evolution in directionally solidified CM247 DS LC during tensile deformation was studied[49], showing that the ductility was not impaired even when the yield strength was at its highest (750C). This level of yield strength was attributed to the presence of intersecting faults, together with partial dislocations which were bounded by antiphase boundary. A high ductility was however retained due to the formation of self-interstitial stacking-faults within γ' precipitates. Appreciable strain-hardening was observed at 750C.

Niobium

Severely plastically deformed Nb-1Zr alloy was produced[50] at room temperature via equal channel angular extrusion at 2.5mm/s and cold rolling to 80% thickness reduction. Following 16 passes, there was a more than 2.5-fold increase in the yield strength together with a ductility which was greater than that of the as-received material. The cold-rolling of processed material led to tensile strengths of above 1GPa. The as-received material had an average grain size of 70μm. Following 8 passes, there was an elongated microstructure with a high dislocation density and an aspect ratio of 5 in which the average long-axis size was 1μm. The grain boundaries were mainly of low-angle type. Grains and sub-grains with a high dislocation density were found following 16 passes. The essentially equiaxed sub-grains had sizes of the order of 200nm. An acceptable mechanical behaviour was exhibited by samples which had been subjected to 16 passes.

Titanium

The microstructures of α/β alloy, Ti-6Al-4V, comprise a combination of 4 phases: the equilibrium aluminium-rich hexagonal close-packed α-phase and the vanadium-rich body-centred cubic β-phase, plus the metastable martensitic phases, orthorhombic α"-phase and the acicular α'-phase, which has the same structure as that of α but is supersaturated with vanadium. The high cooling-rates of additive manufacturing processes such as selective laser-melting result in microstructures which are dominated by α' and which exhibit tensile elongations of less than 8%. The low ductility is attributed to a supposed inherent brittleness of α', and this is commonly avoided by decomposition into α and β. A fully-martensitic titanium alloy which consisted entirely of hexagonal close-packed α'-phase was produced[51] by selective laser-melting. It exhibited a combination of high strength and ductility. A small quantity of body-centred cubic β-phase which formed at large primary α' plates, oriented at about 45° to the tensile direction, was deemed to be responsible for premature fracture, due to α'/β strain-incompatibility. Adjustment of the selected laser-melting parameters led to a β-free microstructure which contained only α'/α' interfaces. This prevented premature fracture. Extensive plastic deformation was then possible, creating randomly oriented nano-α' crystals. This proved that α' is not inherently brittle. The combination of a yield strength of 1150MPa and a tensile elongation of 14 to 15% was possible for a fully martensitic α' Ti-6Al-4V alloy when produced using laser melting. It was essential however to create intersecting ultra-fine α' plates which were free of β-phase.

The effects of the initial microstructure upon the properties Ti-6Al-4V extra-low interstitial alloy subjected to high-pressure torsion at room temperature or 500C were compared[52] for fully lamellar, martensitic, equiaxed and globular. The processing led to appreciable microstructure-refinement, regardless of the initial microstructure, and to considerable increases in hardness, tensile strength and ductility. On the other hand, the elastic modulus of the processed alloy was almost two times lower than that of the alloy in its initial state. The increases depended markedly upon the initial microstructure and the processing temperature. An initial martensitic microstructure improved the hardness, tensile and fracture properties. Processing at 500C was even more effective. Following the latter treatment, material with an initial martensitic microstructure exhibited a hardness of 455MPa, an ultimate tensile strength of 1546MPa, an elongation-to-failure of 18.8% and an elastic modulus of 78.6GPa. Material with an initially fully lamellar microstructure had the lowest elastic modulus (68GPa). An ultra-fine homogeneous $(\alpha + \beta)$ two-phase microstructure which was produced by processing material with an initially martensitic microstructure provided the optimum combination of strength and ductility and strength/elastic-modulus ratio (0.0198).

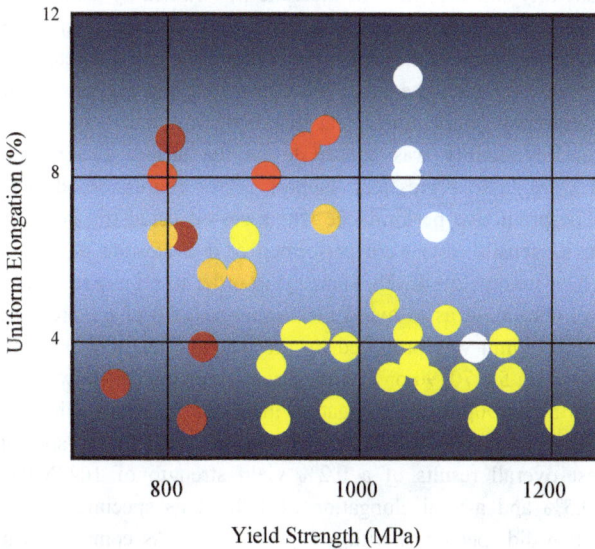

Figure 5. Tensile properties of laser powder bed fused Ti-6Al-4V, compared with other preparation methods. Brown: cast, red: wrought, orange: electron beam melted, yellow: laser powder bed fused, white: present work

It was noted that titanium alloys which are prepared via laser powder-bed fusion tend to exhibit poor ductility because of the presence of acicular martensite in the inherited columnar grains. Heat treatment at about 1075K, to decompose the martensite, improves the ductility but impairs the strength. Artificial intelligence methods were used[53] to optimise both strength and ductility by analysing the effects of the many variables involved in fabrication. Laser powder-bed processed Ti-6Al-4V was used as a test case, and machine learning was used to accelerate the identification of the best sets of parameters required to impart good strength and ductility. The property improvements were the result of the presence of refined prior-β grains, decorated with confined α'-colony precipitates. The improved uniform deformation of the martensite was due to an increased microstructural uniformity resulting from a reduced variant-selection. The machine learning had permitted the rapid and quantitative analysis of the effect of each processing method, and thus the rapid identification of those parameter sets which optimised both ductility and strength. The machine was trained by using parameters such as the distance between adjacent scan paths, together with a data-set based upon 129 similarly fabricated specimens of the same alloy. Other process parameters were the laser power, the laser scan velocity, the layer thickness, line energy density and the post-processing heat treatment temperature, heat treatment time and degree of hot isostatic pressing. The tensile ductility was chosen to be the target. The machine learning algorithms used were lasso regression, support vector regression, decision tree and random forest. The predictive performance was assessed using the Pearson correlation coefficient. Uniaxial tensile tests were performed at a strain-rate of 2.5×10^{-4}/s. The specimens were heat-treated (500C, 2h, vacuum) in order to relieve residual stresses; the temperature being chosen so as to avoid the formation of β-phase precipitates. A line energy density of 0.1J/mm was proposed by the machine-learning prediction, and the hatch spacings were 30, 50, 70, 90 or 110μm. Stress-strain curves showed that, when the hatch spacing was 30 to 90μm, the elongation increased markedly with hatch spacing, according to: elongation(%) = 0.1636 x spacing(μm) - 0.011. A spacing of 90μm imparted the best overall results of a 0.2% yield strength of 1044MPa, a uniform elongation of 10.5% and a total elongation of 15%. This specimen also had shorter martensitic lath than did specimens with a 30μm spacing. As compared with the latter specimens, the prior-β grain in the 90μm specimens was much refined (table 9). The average grain diameter of columnar grains also gradually decreased with hatch spacing. The latter exhibited a critical value at 70 to 90μm: at higher values, prior-β grains were often trapped within track contours, the so-called cage effect. Following machine-learning advice, the parent-phase grains were greatly refined. During loading, the parent-phase boundaries played an important role in stress concentration, given that refined

Materials Research Forum LLC
https://doi.org/10.21741/9781644903230

prior-β grains could provide more interface over which to distribute stresses. The refined prior-β grains also reduced microcrack lengths and promoted post-uniform elongation. During the transformation from β to α', the prior-β grains could furnish 12 variants; tending to produce a weak texture in the fully martensitic microstructure. Specimens with a hatch spacing of 30 and 90μm contained martensite laths of weakly preferred orientation. The overall results strongly suggested that the machine-learning algorithm chose to homogenize the martensite microstructure by adjusting the processing parameters.

Table 9. Diameter of prior-β grain size and hatch-spacing of Ti-6Al-4V prepared by selective laser melting

Hatch Spacing (μm)	Grain Diameter (μm)
25	102.82
45	81.96
65	74.61
85	66.45
105	56.00

Zirconium

Metallic glasses are notorious for their high strengths but poor plasticity, and are therefore excellent targets for trying out methods by which to evade the strength-ductility paradox. Unlike crystalline alloys, plastic deformation in metallic glasses at room temperature arises mainly from stress-driven local cooperative motion in shear-transformation zones which generate shear-bands. Assuming a 2-step route, involving so-called mechanical rejuvenation and cyclic stress-loading, in the case of a model Zr_2Cu metallic glass the deformation behavior was analysed[54] using molecular dynamics simulations. Simulated uniaxial compressive stress-strain curves revealed that there were marked variations in the UTS, YS and elastic strain limit. In general, the rate of strain-increase began to slow when the elastic limit was passed. At higher energy states of the glass, characterized by an increased compressive plasticity, a so-called mechanical rejuvenation was due to internal stress accumulation and an anomalous increase in free volume. Cyclic stress loading during rejuvenation removed loosely-packed regions and produced a more stable state of lower energy, thus improving structural stability with a concomitant improvement in the elastic strain limit and strength. Theoretical modelling explained a strength-ductility compromise in terms of a delay in shear-transformation zone behaviour which governed the generation and intersection of multiple shear-bands. A 2-step process in which soft regions aggregated into a concentrated mass of higher potential energy, followed by dissipation of that mass into randomly dispersed soft regions was the key to evading the strength-ductility paradox. Repetitive cyclic loading and unloading could provoke a transition from strain-softening to strain-hardening, indicating that an improvement in yield stress and elastic strain limit could be obtained without impairing plasticity.

Ferrous Alloys

It was recalled that the strength of metallic materials can be increased by pinning the dislocations, which mediate plastic deformation, by using nanoscale obstacles. On the other hand, simultaneously obtaining ultra-high strength and ductility was very difficult in conventional metallic materials. An iron-based alloy with increased ductility was produced here[55], in which the strength was twice that of the upper limit of conventional alloys. This was done by exploiting the paradoxical concept of lattice softening. This in turn was achieved by tailoring atomic arrangements having a specific electronic structure. A nanograined structure was then produced via severe plastic deformation.

In a review[56] of the ductility of nanostructured and ultrafine-grained iron produced by extreme plastic deformation, it was noted that powder milling and hot-consolidation led to a high mechanical strength but poor ductility. Improvements in the process could increase the total plastic strain of nanostructured iron, and the creation of bimodal structures promoted strain-hardening and a more uniform deformation. In the case of steel however, the hardness of the milled powder and the role played by carbon atoms in ferrite grains made it more difficult to improve the ductility of nanostructured material.

Attempts have also been made[57] to avoid the compromise between strength and ductility by using means other than severe plastic deformation. Quenching and tempering variants, such as quench – partitioning – tempering, increase strength while there are various methods for meanwhile increasing the ductility by modifying the deformation mechanism. Such methods include shear-band induced plasticity, transformation-induced plasticity and twinning-induced plasticity. A then-new approach, which was hoped to yield MPa% values of above 65000, involved the formation of a microstructure consisting of two distinct microconstituent structures which played distinct roles in deformation and strengthening, but acted synergistically so as to develop novel combinations of strength and ductility. The deformation response involved sequential stress-activated processes which led to a combination of dislocation-dominated mechanisms, with the formation of high-density dislocation networks, twin/defect annihilation, phase transformation, matrix-phase nanoscale-refinement and nanoprecipitation.

The properties of ultrafine-grained steels with a grain size much smaller than 1μm were studied[58]. They attained strengths which were to 2 to 4 times those of coarse-grained equivalents. The uniform tensile elongation was however limited to a few percent when the materials had a single-phase ultrafine-grained structure, and this was attributed to early plastic instability. A titanium-added ultralow-carbon interstitial-free steel was subjected to high-deformation accumulative roll-bonding in order to obtain such an ultrafine-grained microstructure. Lengths of 1mm steel sheet were stacked, spot-welded at each end, held at 500C for 600s and roll-bonded by 50% one-pass rolling. This process was repeated up to 7 times. The resultant sheets were annealed (400 to 800C, 0.5h) in order to vary the grain size. Sheets of SS400 with a fully martensitic structure were prepared by austenitization and water-quenching. They were cold-rolled to 50% total reduction, and annealed (200 to 700C, 0.5h). Ultrafine-grained structures comprising ferrite with a grain size of 100nm and uniformly dispersed nano-carbides could be formed. The mechanical properties were measured at room temperature, using an initial strain-rate of 0.00084/s. The strength of interstitial-free sheet was markedly increased by a single cycle of accumulative roll-bonding, while the tensile elongation greatly decreased. The strength increased with increasing roll-bonding cycles and the tensile strength attained 900MPa after 7 cycles; 3.2 times that (280MPa) of the original sheet. The elongation was however almost the same after 1 cycle, and limited to a few percent. These materials had an ultrafine-grained microstructure which was elongated along the rolling direction. It was noted that, if the strain-hardening could be increased in some way, plastic instability would be delayed. One method would be to disperse a fine second phase within the fine ferrite matrix. In the low-carbon SS400, there was an ultrafine-grained structure with nanometre-sized carbides uniformly precipitated within the fine matrix. This multi-phase steel managed to offer both a high strength and adequate ductility. Strain-hardening was clearly enhanced and a specimen annealed at 550C had a tensile strength of 850MPa, a uniform elongation of 10% and a total elongation of 10%. The usual tensile strength of the steel was 400MPa. This showed that the dispersal of fine precipitates was effective in achieving both high strength and good ductility.

Dual-phase steels, with microstructures comprising soft ferrite and hard martensite phases, offer good strain-hardening, high strength and high elongations. Two types of structure, with networks of martensite or isolated martensite phases, were produced[59] in a

Materials Research Forum LLC
https://doi.org/10.21741/9781644903230

low-carbon steel (Fe-0.087C-0.79Si-1.77Mn-0.02Cr-0.01wt%Mo) in order to clarify the reason for such properties. The samples with networks of martensite offered the better strength-ductility balance. Microscopic analysis of equivalent areas showed that the strain distributions within the dual-phase microstructures were non-uniform and that the plastic strain was localized to within soft ferrite grains. The strain eventually propagated across the martensite however. In detail, sheets (1mm-thick) of the steel were austenitized (950C, 0.5h) and air-cooled to room temperature so as to obtain a microstructure comprising ferrite and a small amount of pearlite. Sheets with the ferrite-based microstructure were then annealed (760 to 840C, 0.5h) in the ferrite/austenite 2-phase region, followed by water-quenching, in order to produce microstructures with the networked martensite. Other sheets were austenitized under the above conditions, cooled to between 720 and 820C, and annealed at the holding temperature for 0.5h, followed by water-quenching to obtain microstructures with the isolated martensite. With increasing intercritical annealing temperature, the volume fraction of austenite increases and volume fraction of martensite in specimens following water-quenching increased. The grain sizes of ferrite and martensite ranged from 2 to 5μm. The volume fractions of martensite were similar in both types of microstructure. The grain size of martensite (the grain size of the prior austenite) was greater than that the network microstructures. For both networked and isolated martensite the tensile strength decreased with increasing total elongation, reflecting the usual paradox. As noted above, specimens with networked martensite offered a better strength-ductility balance than did structures with isolated martensite. It was noted that the martensite in the networked structure was more deformed than the martensite in the isolated structure. The distributions of local strain in the ferrite and martensite phases, in the networked structure, were broader than those in the isolated structure, suggesting that the degree of strain-partitioning between ferrite and martensite decreased when the martensite was a connected by a network. That is, the deformation becomes more homogeneous in the networked structure than in the isolated structure, thus contributing to a better strength-ductility balance.

Alternating stacks of cold-rolled and annealed interstitial-free steel were subjected to hot-compression followed by cold-forging with the object of improving the strength without an excessive loss of ductility[60]. The initial materials were 1mm-thick titanium-added commercial cold-rolled (78% reduction) steel sheet. Some sheets were annealed (780C) to produce an average grain size of about 20μm. Layers of cold-rolled and annealed material were stacked in an alternating sequence of 11 layers along the original rolling direction, and joined by compression at 500C in vacuum. The hot-compressed material was then air-cooled and cold-forged at room temperature to 1.6mm. The deformation strains in the cold-rolled and annealed layers were then 96.8% and 85.5%, respectively.

The cold-forged material was then annealed (625C, 0.5h or 1h) and tensile-tested along the original rolling direction. In the annealed layers, there were mainly coarse recrystallized grains plus fine-grained deformed lamellae which were almost parallel to the rolling direction. In the cold-rolled layers, the microstructure comprised a fine-grained deformed matrix plus coarse grains. The average grain sizes of the small grains and coarse grains were 1.6μm and 11.1μm, respectively. The annealed material had the highest uniform elongation, and a yield strength of some 150MPa. Heavily deformed material exhibited barely any uniform elongation. The strength of the cold-forged material was greater than that of cold-rolled material, due to deformation strain. Following heat treatment (625C, 0.5h), the uniform elongation was about 30%, and the yield strength was twice that of the annealed material. Annealing for 1h led to a decrease in strength, but it was still greater than that of material prepared by accumulative roll-bonding, to a similar strain, followed by annealing. The average size of the fine grains in the 1h-annealed material was about 1.6μm. Overall, the combination of strength and ductility in the present work was better than previously reported data for homogeneous ultra-fine grained interstitial-free steel prepared by equal-channel angular pressing/extrusion, accumulative roll-bonding or asymmetric rolling. It was suggested that the ultra-fine grains led to a high strength while the large grains controlled the tensile elongation due to the interfaces between layers and lamellae having differing grain sizes and orientations within the layers. A high strength and good ductility could thus be achieved in the present steel, and the key parameters were the layer thickness, the stacking sequence, welding method, deformation strains and heat treatment.

Samples of pearlitic steel (0.76C-0.35Si-1Mn-0.017P-0.014wt%S) were subjected to high-pressure torsion[61]. The size of the randomly oriented pearlite colonies in the undeformed material ranged from 10 to 20μm and, within the colonies, the cementite-lamellae spacing was about 200nm. With increasing strain, the lamellae became aligned parallel to the shear plane and the lamellar spacing decreased. Unfavourably-aligned lamellae were heavily bent and broke into smaller pieces, which then aligned during further shearing. At strains greater than 8, the lamellae were almost fully aligned while, at a strain of 16, the lamellar spacing was reduced to 15 to 20nm. There was an increase in hardness with increasing strain, and it increased from 270HV in the undeformed state to 770HV at a strain of 16. In some samples, the crack plane was parallel to the shear plane and the propagation direction was tangential to the twisted sample or parallel to the shear direction. In other samples, the crack plane was perpendicular to the shear plane and crack growth occurred along the axial direction of the twisted sample and perpendicular to the shear plane. In a third type of sample, the crack plane was perpendicular to the shear plane but crack propagation occurred in the radial direction of twisted samples.

With increasing strain, a marked anisotropy of the fracture toughness appeared. In the first of the above cases, the fracture toughness decreased very quickly at small shear strains and attained an almost constant low value at higher strains. In the last of the above cases, the fracture toughness increased at low strains but decreased at higher strains and approached the values found ($40MPam^{1/2}$) for undeformed material. The fracture toughness in the second of the above cases was between the values for the other cases, but the crack immediately deviated into the shear plane; perpendicularly to the initial crack plane and following the direction of lowest toughness. The overall results were relevant to the processing of cold-worked pearlitic steels and to the strengths and ductilities which were obtainable in such materials. It is assumed that the strength is the sum of the contributions arising from the Hall-Petch and Orowan mechanisms, from dislocation-hardening and from solution-hardening. In typical loading directions, the fracture resistance is high while, in directions where the stresses are low or compressive stresses are present, the fracture toughness is very low. Cracks at the microscale are unavoidable during processing and, during the tensile loading of sheets or wires, such cracks encounter a high fracture-resistance due to deflection or delamination. It was suggested that a decrease in fracture toughness and a decrease in the anisotropy of cold-drawn hypereutectoid steel wires with strengths exceeding 6.5GPa might explain why it became increasingly difficult to deform such wires further. During micro-compression there was a marked tendency for the formation of a single so-called macro shear-band during loading of the lamellar microstructure in the normal direction; compression perpendicular to the lamellar plane. A result of the anisotropy of shear localization appeared to be that, during rolling, the occurrence of macro shear-bands at high strains was much more likely than during wire-drawing. Shear deformation during high-pressure torsion involved relatively homogeneous shearing, and corresponded to the deformation of inclined micro-pillars which experienced the lowest flow-stress. During high-pressure torsion the macro-shear direction then paralleled the direction with the lowest flow stress, such that shearing in another direction would require higher shear stresses. The anisotropy thus stabilized homogeneous shear deformation.

The behaviour of hot-formed 22MnB5 steel with an aluminium-silicon coating was studied[62] with the aim of increasing both the strength and ductility so as to further both vehicle lightweighting and increased energy absorption during vehicular collisions. The higher strength (1900MPa) of the MBW-K1900 variant permitted the production of lighter parts than did 22MnB5. Better crash performance required high strength, without the material cracking during folding or bending. This was the motivation for Tribond 1200 and Tribond 1400, composites in which 2 outer layers of ductile 500MPa steel were bonded to a 1500MPa steel core layer: that is, MBW500 and MBW1500, respectively.

The aluminium-silicon coating permitted Tribond 1200/1400 to be processed in the same manner as similarly coated 22MnB5. When compared with 22MnB5, the MBW-K 1900 offered a further increase in the hardness of the martensite following hot stamping. This was due to the higher carbon content, of up to 0.38wt%. The maximum manganese content remained at 1.4wt%, and up to 0.005wt% of boron was used to increase the time for microstructure transformation. Due to the great hardness (0540 to 620HV) of MBW-K 1900, some stress-relief of the martensite by tempering was required. The effective annealing (170C, 20min) during processing led to an increase in yield strength of 100 to 160MPa while the ultimate tensile strength was reduced by 50 to 100MPa. This was attributed to a greatly increased diffusion of carbon to dislocations, thus hindering glide on the planes of the lattice. By using so-called tailored tempering, the properties of local areas could be adjusted. Even if the properties (e.g. strength and ductility) were contrary, the Tribond family permitted standard steel grades for hot- or cold-forming to be combined into a composite structure so as to improve more than one property simultaneously. In Tribond 1200 and Tribond 1400, the mechanical properties could be adjusted by varying the relative thicknesses of the layers. Tribond 1200 was aimed at ductility and each outer layer made up 20% of the total sheet thickness. Tribond 1400, with an outer layer thickness of 10%, was aimed mainly at high strength. For the combined materials, the ultimate tensile strength was between that of Tribond 1200 (1100MPa) and that of Tribond 1400 (1300MPa). The ductility of the clad material could not be measured properly by a tensile test. However, the MBW 1500 reached a bending angle greater than 50°, Tribond 1400 bettered 75° and Tribond 1200 was greater than 130° for 1.5mm sheet following austenitization (920C, 360s) and hot stamping.

The strength and ductility of normalized (900C, 30min) JIS FCD 700 steel (3.68C-2.41Si-0.43wt%Cu) were studied[63] following austempering (315C, 19, 30, 50 or 77min). The austempering time affected the properties, according to the nature of the retained phases in the microstructure (table 10).

Table 10. Hardness of phases in normalized and austempered JIS FCD 700

Treatment	Phase	Hardness (HRC)
Normalized	nodular graphite	47.5
Normalized	ferrite	170.3
Normalized	pearlite	252.0
austempered, 19min	nodular graphite	47.7
austempered, 19min	martensite	792.2
austempered, 30min	nodular graphite	46.4
austempered, 30min	martensite	651.3
austempered, 30min	acicular ferrite	328.9
austempered, 50min	nodular graphite	47.7
austempered, 50min	martensite	638.3
austempered, 50min	acicular ferrite	322.0
austempered, 77min	nodular graphite	47.9
austempered, 77min	martensite	628.8
austempered, 77min	acicular ferrite	315.1

During cooling to room temperature, carbon atoms diffused from the austenite to the nodular graphite and the solubility of carbon in austenite began to decrease at this point, creating carbon-depleted zones … and thus effective ferrite-nucleation zones. Pearlite nucleated at the ferrite/austenite boundaries as the eutectoid temperature was reached. Following austempering for 19min, the microstructure consisted of martensite and nodular graphite. This treatment time was too short for residual austenite to decompose into acicular ferrite and high-carbon austenite, and the residual austenite transformed into martensite upon cooling to room temperature. The microstructure after austempering for 30min consisted of nodular graphite, martensite and acicular ferrite. The latter was present due to the quite high silicon content, which increased the carbon activity in ferrite and prevented cementite formation. Austempering for 50 or 77min also produced nodular graphite, martensite and acicular ferrite. Normalizing led to the lowest tensile and yield strengths, due to the presence of ferrite and pearlite. Austempering for 19min increased the strength, due to the presence of martensite, but led to brittleness. Austempering for 30min increased the tensile and yield strengths because the acicular ferrite absorbed some strain and reduced the latter brittleness. Austempering for 50min also increased the tensile and yield strengths, and this was attributed to the formation of higher amounts of retained austenite, while decreased amounts of martensite and retained austenite were blamed for the reduced tensile and yield strengths following 77min of austempering. The normalized material exhibited a reasonable elongation because of the ferrite and pearlite content. Austempering for 19min led to the lowest elongation because martensite predominated in the microstructure. Austempering for 30min led to the highest elongation, due to the formation of acicular ferrite. Austempering for 50min led to a lower elongation however, presumably because a higher amount of retained austenite was available to transform into martensite during plastic deformation. Austempering for 77min again caused the elongation to decrease, due to the formation of carbides. The hardness levels generally mirrored the strengths, and for the same reasons. The greatest hardness was recorded following 19min of austempering time, due to the martensite content. Overall (table 11), more retained austenite led to a higher abrasive-wear resistance and strength, and acicular ferrite led to a lower hardness. An austempering time of 50min was optimum and produced an ultimate tensile strength of 1058MPa, a yield strength of 859MPa, a ductility of 4.08%, a hardness of 37HRC and an abrasive-wear resistance of $2.1 \times 10^{-7}mm^3/mm$ under 6.32kg loading. This combination was due mainly to the large amounts of retained austenite.

Table 11. Summary of heat-treatment effects upon JIS FCD 700

Property	Best	Worst
UTS	austempering, 50min	normalizing
YS	austempering, 50min	normalizing
Elongation	austempering, 30min	austempering, 19min
Hardness	austempering, 19min	normalizing
Wear Resistance	austempering, 50min	normalizing

An heterogeneous twinning-induced plasticity steel which contained alternating columnar-grain and equiaxed-grain domains was prepared[64] by means of pre-straining, partial recrystallization and directional solidification. When compared with equiaxed-grain material (table 12), the heterogeneous material exhibited a better tensile strength-elongation combination, with a yield strength of 263MPa, an ultimate tensile strength of 573MPa and a total elongation of 101.5%. The constraints of the structure controlled the deformation of the equiaxed- and columnar-grain domains at different stages. The heterogeneous material exhibited stable continuous plastic deformation, due to the release of localized high strain or stress, delayed plastic instability and restrained crack-propagation. The coarse unidirectional columnar grains contributed to the ductility, while the fine randomly-oriented equiaxed grains contributed to the strength. The samples differed with regard to their work-hardening rates at low to medium strains. Here the equiaxed material exhibited an essentially unchanged work-hardening rate with increasing true strain and the highest rate corresponded to a true strain of 0.42. The heterogeneous material exhibited a continuously increasing work-hardening rate.

Table 12. Tensile properties of twinning-induced plasticity steels

Steel	YS (MPa)	UTS (MPa)	e_t (%)
Directional	213	503	102.0
Equiaxed	217	561	82.8
heterogeneous	263	573	101.5

Equal-channel angular extrusion is able to improve moderately the strength of ferritic T91 steels. Various strategies were explored[65] in order to improve the strength further, while maintaining ductility. Thermomechanical treatments involving combinations of angular extrusion, water-quenching and tempering (table 13) could produce a so-called ductile martensite. The ductile martensite had a much greater strength than that of fine-grained ferrite. It was possible to tailor the microstructure and properties, using equal-channel angular extrusion, so as to obtain an improved combination of strength and ductility. Ausforming or austenitizing of the steel at 900C or above produced mainly martensite and the transformed martensite, combined with auto-tempering, furnished materials having good ductility. Bainitic ferrite was also formed during the transformation. Transitional carbides were found in the water-quenched steel. The strengthening of the processed steel largely involved the Hall-Petch effect plus dislocation-strengthening and solid-solution strengthening. Three distinct zones were identified (figures 6 and 7), based upon the primary phase following processing. In the lower ferrite zone, angular extrusion of ferrite-phase steel produced fine ferrite. In general, a higher strength was associated with a lower ductility. The highest strength attained was 1000MPa, with 2% of uniform elongation. In the upper martensite zone, including water-quenching from 900 to 1200C, angular extrusion produced a martensite which offered a combination of high strength and reasonable ductility. The yield strength could exceed 1600MPa, with about 1.5% of uniform elongation. Heat-treatment increased the uniform elongation to about 4%, with a yield strength of 1400MPa. In the ferrite/martensite zone, the steel was processed using mainly water-quenching from 800C and tempering at 600C. Water-quenching from 800 to 1200C led to a ductile martensite which exhibited several percent of uniform elongation, because of auto-tempering. The formation of ferrite reduced the yield strength by about 150MPa. When austenitizing was carried out above 900C, water-quenching led to the production of steel composed mainly of martensite. The as-quenched martensite had a yield strength of about 1200MPa and a uniform elongation of 6 to 8%. The processing thus made it possible to adjust the mechanical properties over a broad range. The retention of good ductility was attributed to the low carbon content and to the auto-tempering of martensite during cooling. The strength and ductility of martensitic and ferritic material exhibited the paradox, but ductile martensite could push the envelope.

Table 13. Microstructures and properties of T91 steel following austenitizing (1h) and water-quenching

T(C)	Initial Phases	After Quench	YS(MPa)	e_u(%)
800	γ, undissolved α, undissolved carbides	$α'_M$, $α_A$, $α_P$, carbides	972	7.2
900	γ, undissolved carbides	$α'_M$, $α_P$, carbides	1127	
1000	γ, undissolved carbides	$α'_M$, $α_P$, carbides	1207	
1100	coarse-grained γ	$α'_M$, $α_B$, carbides	1212	6.9
1200	coarse-grained γ	$α'_M$, $α_B$, carbides	1156	6.8

$α_A$: grain-boundary ferrite, $α'_M$: martensite, $α_P$: ferrite

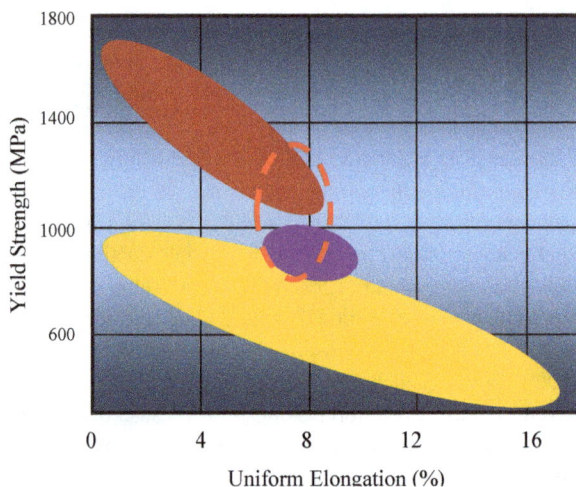

Figure 6 Uniform elongation versus yield strength for T91 steel with various microstructures following extreme plastic deformation. Dark red: martensite, purple: martensite/ferrite, orange: ferrite. Light red: water-quenched

In later work on the steel[66] it was noted that 4 main strengthening mechanisms contribute to the yield strength of ferritic/martensitic steels: solid-solution strengthening, precipitation-hardening, dislocation-strengthening, boundary-strengthening. The latter factor included lath, packet and inherited austenite grain boundaries. Boundary strengthening arising from fine martensite, solid-solution strengthening arising from dissolved carbon atoms and precipitation-hardening arising from θ-carbides were studied here. Sub-grain boundary-strengthening played a major role with regard to the yield strength of as-received, water-quenched and partially tempered samples. The finer martensitic blocks and laths in partially tempered samples imparted a higher strength. This contribution could however be overestimated because the length of martensitic blocks and laths was overlooked. The effect of substitutional solute atoms upon strengthening was much less than that of interstitial atoms, so that carbon and nitrogen were most effective in improving the strength of ferrite; the contribution being closely related to the carbon or nitrogen concentration. Solid-solution strengthening arising from carbon in the T91 steel was strong in some partially-tempered samples, and in water-quenched samples, but was weak in as-received and other partially-tempered samples, due to the precipitation of $M_{23}C_6$ and MX particles. In ferritic/martensitic steels with a low carbon content, the dislocation density within the martensitic laths was typically 10^{14} to $10^{15}/m^2$ following quenching. After tempering, it was within the same range. The dislocation density for fully-tempered T91 steel was $1.7 \times 10^{14}/m^2$, leading to a strength of 290MPa for as-received material. The dislocation distributions ranged from tangles to recovered cell structures in the various samples. The degree of dislocation-strengthening in water-quenched and samples partially tempered at 500C was expected to be similar, but somewhat greater than that in as-received material, due to a relatively higher dislocation density. Precipitation-hardening in as-received samples (table 14) was not notable in improving the yield strength, due to the relatively coarse $M_{23}C_6$ and MX particles and their large spacing. The hardening in water-quenched samples was relatively weak because they were strengthened more by carbon in solid solution and by lath boundaries. Due to the formation of bainitic ferrite grains in samples partially tempered at 500C, the lath-widths were quite small and bainitic ferrite also strengthened via ultra-fine martensite and dispersed transition carbides. This all made the partially tempered material stronger than the water-quenched material (table 15). Sub-grain strengthening was the main strengthening mechanism in samples partially tempered at 500C, but the values were slight overestimates, due to elongated martensitic laths. Solid-solution strengthening was a secondary effect in the partially tempered material, and similar to its effect in water-quenched material. The effect of dislocation-strengthening was similar for all of the materials. Although precipitation-hardening was small, it led to a higher yield strength of partially tempered samples. A greater uniform elongation of partially tempered material as compared with that of water-quenched samples was attributed to barrier effects arising from relatively recovered dislocations, a high density of precipitates and relatively fine martensitic laths. It was concluded that partial tempering at the intermediate temperature of 500C produced samples having a high yield strength (~1.4GPa) and a relatively good uniform elongation (~5.5%).

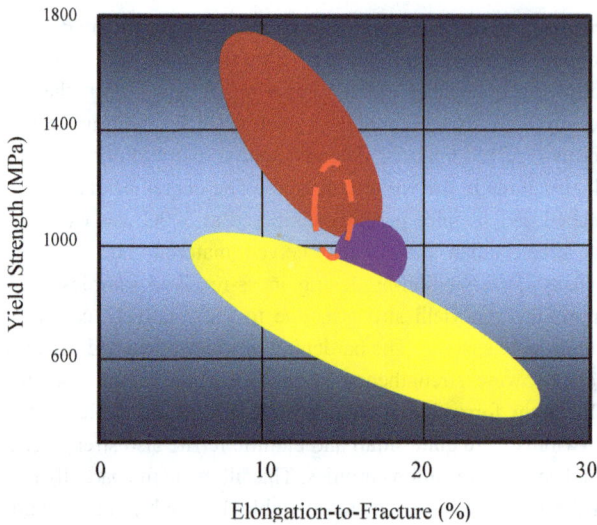

Figure 7. Elongation-to-failure versus yield strength for T91 steel with various microstructures following extreme plastic deformation. Dark red: martensite, purple: martensite/ferrite, yellow: ferrite. Light red: water-quenched

Table 14. Tensile properties of as-received, quenched or tempered T91 steel

Treatment	YS(MPa)	UTS(MPa)	e_u(%)
as-received	675	817	8.2
water-quenched	1190	1410	3.8
partially tempered (300C)	1070	1348	5.3
partially tempered (500C)	1370	1506	5.5
partially tempered (600C)	1090	1263	6.5
partially tempered (700C)	880	1011	4.6

Table 15. Strengthening contributions to T91 steel

Condition	σ_{SG}(MPa)	σ_{SS}(MPa)	σ_{P}(MPa)	σ_{D}(MPa)
as-received	360	-	70	290
water-quenched	690	500	-	290
partially tempered (500C)	920	500	200	290

σ_{SG}: sub-grain boundaries, σ_{SS}: solid solution, σ_{P}: precipitates, σ_{D}: dislocations

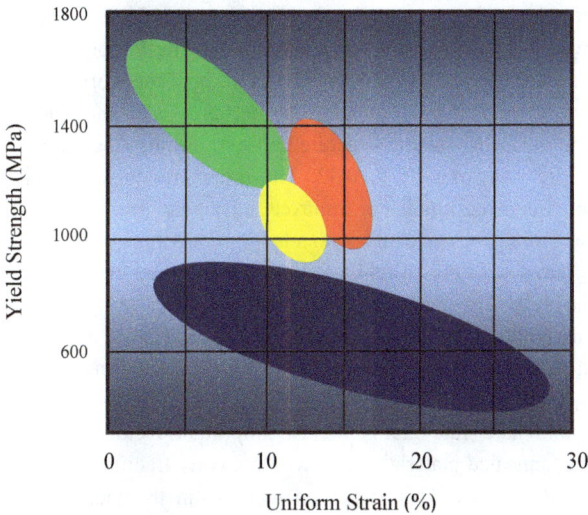

Figure 8. Yield strength versus uniform elongation for various T91 steel conditions, following extreme plastic deformation. Green: α', red: $\alpha_B+\alpha'_{PT}$, yellow: α+α', blue: α. PT: partially tempered, B: bainitic.

The properties of Fe-30Mn-10Al-1C low-density steel following cold-rolling (80% thickness reduction) and annealing were investigated[67], showing that nano-sized κ-carbides precipitated unevenly during 500C annealing (180s), pinning austenite grain boundaries and limiting the growth of austenite grains during later annealing at 900C. The 2-step annealing produced a fine heterogeneous grain structure which led to an improvement of some 30MPa in the yield strength, to 752MPa, while hardly affecting the uniform elongation. Unlike 1-step annealing (900C, 180s), the 2-step annealing imparted a strength x ductility value of 58GPa% and an impact toughness of 128J/cm² at -40C. Both types of sample yielded continuously in tension, with no appreciable yielding terrace. The yield strength of 1-step annealed material was 721MPa, the ultimate tensile strength was 1065MPa and the elongation was 53.6%. The 2-step annealed material had a yield strength of 752MPa, an ultimate tensile strength of 1076MPa and an elongation of 53.5%. Both materials exhibited similar work-hardening behaviours, with a continuous decrease due mainly to strain localization. Three stages of work-hardening were associated with slip-band refinement. In the first stage, the accumulation of dislocations on the slip surface saturated, slip-bands penetrated the entire grain and the work-hardening rate decreased rapidly at true strains of 0.004 to 0.007. In the second stage, with increasing strain, new dislocation sources accommodated further plastic deformation and new slip-bands caused the spacing between slip-bands, and the work-hardening rate, decreased slowly from a strain of 0.007 to 0.01. In the third stage, the work-hardening rate rapidly decreased and the sample fractured after reaching the limit of plastic deformation from a true stain of 0.01 to 0.32. The microstructures of both types of material consisted of fully recrystallised equiaxed austenitic grains, but the 2-step annealed structure was more heterogeneous, with a laminar distribution of coarse and fine grains. The average grain size of the austenite following 1-step and 2-step annealing was 2.9 and 2.3μm, respectively. The main reason for the finer austenite grain size in 2-step annealed material was that a large amount of κ-carbide precipitated after annealing at 500C. Grain-refinement led to a higher work-hardening rate and an improved yield strength, in the low-strain state, of 2-step annealed material. The strain distribution between soft coarse and hard fine grains was an important factor in improving the properties of the 2-step annealed material while avoiding early fracture via severe strain-concentration in the finer grains. This led to an increase in the yield strength while retaining a good plasticity. The difference in the yield strengths of the samples was due mainly to grain boundary strengthening. That contribution for 1-step annealed material was about 271MPa and that for the 2-step annealed material was about 304MPa. The difference of 33MPa was very close to the experimentally measured difference in the yield strengths.

Samples of fine-grained Fe-34.5Mn-0.04C steel with recrystallized grain-sizes of 3.8 to 2.0μm were prepared[68] by 90% thickness-reduction cold-rolling, followed by annealing (650 or 800C). A simultaneous increase in strength and ductility resulted from grain refinement, such that the best combination of strength and ductility was found for samples having the finest grain size. Tensile tests in the rolling direction were carried out at room temperature using a strain-rate of 0.001/s. A banded structure, with the banding parallel to the rolling direction, was present in which the band-width varied from 1 to 10μm. Some of the bands had wavy features which were associated with localized shear deformation. A fine deformed-lamellar structure, with the boundaries approximately parallel to the rolling direction formed. The spacing between the lamellar boundaries was much finer than the band-widths seen in optical images. The lamellar boundary spacings along the normal direction ranged from a few nanometres to about 150nm, with an average of about 40nm. In addition to the deformed lamellar structure, nanotwin-bundles and localized shear bands were found with an areal fraction of about 18%. The 90% cold-rolled material was annealed in various ways so as to produce 4 different grain sizes (table 16). In samples annealed at 650C for 300s the grain structure was uniform, apart from some 2% of non-recrystallized deformed material. Samples which were annealed under other conditions were fully recrystallized. It appeared that a grain size of 2.0μm was the minimum which could be obtained by annealing the 90% cold-rolled steel. Grain refinement from 3.8 to some 2.0μm resulted in the noted simultaneous increase in strength and ductility, but also in a transition from continuous yielding (3.8μm grain size) to undesirable discontinuous yielding for finer grain sizes. The change in the yielding behaviour was associated with a marked increase in the yield strength from 256 to 405MPa as the grain size varied from 3.8 to 2.2μm. The lower yield strength was particularly sensitive to the change in grain size, with a decrease from 2.2 to 2.0μm leading to an increase from 405 to 466MPa. In order to suppress the undesirable discontinuous yielding, annealed material with a 2.0μm grain size was subjected to 2% thickness reduction by cold rolling. This resulted in the yielding behaviour becoming continuous, together with a further increase in the yield strength.

Table 16. Mean grain size and tensile properties of Fe-34.5Mn-0.04C steel

Annealing	Grain Size(μm)	YS(MPa)	UTS(MPa)	e_u(%)	e_t(%)
650C, 5min	2.0	466	640	32.6	39.6
650C, 10min	2.1	444	635	35.6	39.4
650C, 30min	2.2	405	620	37.5	38.8
800C, 60min	3.8	256*	550	32.5	34.1

*0.2% off-set, e_u: uniform elongation, e_t: total elongation

A single-step process was used[69] to create bimodal grain structures in the SS316L austenitic stainless steel. The bimodal structure consisted of fine (less than 500nm) martensite grains, sandwiched between coarse (circa 10μm) austenite grains. The dual-phase bimodal structure exhibited a yield strength of about 620MPa and a uniform tensile ductility of about 35%. The tensile tests were performed at room temperature using a strain-rate of 0.001/s. These results were attributed to the fact that the dual-phase bimodal grain structure delayed the onset of plastic instability; resulting in a higher strength and a higher uniform elongation and work-hardening rate. In the process, submerged friction-stir processing was used to tailor the surface properties with the aid of a tungsten carbide tool with a 12mm shoulder-diameter. The rotation-speed was 1800rpm, the plunge-depth was 0.4mm and the tool was traversed in the longitudinal direction at 20mm/min while submerged in a 50:50 mixture of distilled water and ethanol at 0C. The tool could be rotated at a particular location for 0.25h in order to produce localized straining. Traversing the tool along the specimen produced an ultra-fine grain structure. Rotating the tool in one location produced the bimodal structure of fine grains in a matrix of coarse grains. The depth of processed region in both cases was nearly 300μm. The as-received steel exhibited a wide range of grain sizes, with an average of 22μm. The ultrafine-grained material had an average grain size of nearly 0.9μm. The average grain size of the bimodal material was nearly 3.5μm. The volume-fractions of martensite were nearly 30% and 8% in the bimodal and ultrafine-grained material, respectively. The as-received steel was predominantly austenite. The ultrafine-grained material consisted of a large volume-fraction of elongated deformation-bands separated by thin boundaries. There was a high density of dislocations within the deformation bands. The average band width was 40 to 50nm. The bimodal material had fine deformation-bands and sub-grain regions of very high and low dislocation density. The as-received steel had a yield strength of 300MPa, with 50% uniform elongation. The yield strength of the ultrafine-grained material was increased to 450MPa, with a reduced uniform elongation of 30%. The variation in work-hardening rate, as a function of true strain, was similar for all of the materials and involved 3 distinct stages. An initial high rate of work-hardening was followed by a reduced and nearly steady-state stage, with a minimum work-hardening rate at higher strains. All of the tensile specimens exhibited similar and extensive dimple formation; indicating an appreciable plastic deformation up to failure. The friction-treatment led to a situation in which the cold-worked structure was replaced by recrystallized grains. The recrystallization in these materials involved the generation of a new grain structure via distinct nucleation and growth (so-called discontinuous dynamic recrystallization). During recrystallization, the nucleated grains formed a band of recrystallized grains in a necklace-like pattern, eventually leading to a fully recrystallized structure. Energy stored

Materials Research Forum LLC
https://doi.org/10.21741/9781644903230

in the structure during deformation was the driving force for recrystallization. The bimodal structure was attributed to a mechanism in which the external strain causes sliding/migration of the parent grain boundaries. The latter act as nucleation sites for the development of new stress-free grains. The higher the strain-rate, the lower is the nucleated grain size. With decreasing processing time at a fixed location, the bimodal structure became coarser: the average size of austenite grains increased to nearly 15µm while the fraction of fine martensite grains was reduced. During stationary processing, the local strain caused pile-ups of dislocations along the grain boundaries. Those dislocations then formed slip-bands, with twins and faults which acted as martensite nucleation-sites. All of these interacting phenomena led to the dual-phase bimodal grain structure. The bimodal material exhibited a greater work-hardening, a higher yield strength and a greater elongation than that of ultrafine-grained material. The fine martensite grains, embedded in the coarse austenite matrix, could also be treated as a dispersion-strengthened material. In such materials, the work-hardening rate depends mainly upon the average dislocation density around particles. The combined high yield strength and work-hardening rate of bimodal material was attributed to the higher volume fraction of finer martensite grains, while the larger elongation was attributed to high dislocation-densities at the hard non-deformable martensite grains and the high dislocation storage by coarse austenite grains. So, in essence, the high fraction of fine martensite grains within the coarse austenite delayed the onset of plastic instability and led to appreciable work-hardening, thus providing a means for evading the strength-ductility paradox.

Excellent corrosion resistance was obtained[70] for a bimodal stainless steel by using a single-step process. The microstructure consisted of fine (200 to 400nm) martensite within a matrix of coarse (10µm) austenitic grains. Such bimodal microstructures are attractive because they simultaneously impart a high strength and good ductility, thus evading the strength-ductility paradox. It is a bonus that such steel also exhibits a much lower corrosion rate (0.001mm/year in 3.5wt%NaCl solution) than that (0.088mm/year) of commercial 316L stainless steel.

Coarse-grained austenitic 304 stainless steel was reduced, post-homogenisation (1050C, 1h) and water-quenching, to the nano-grained level by 8 passes of equal-channel angular pressing[71]. The nano-grain microstructure, with its high dislocation density, led to a higher strength but a lower ductility. Recrystallization of the nano-grained material involved 3 distinct stages: a recovery stage below 650C, a recrystallization stage ranging from 650C to 800C and a grain-growth stage at above 800C. Annealing (750C, 0.5h) produced a bimodal microstructure which consisted of 62vol% of ultra-fine grains with an average size 350nm and 38vol% of coarser grains with an average size of 1.4µm. This structure imparted a yield strength of 725MPa, a tensile strength of 930MPa, a uniform elongation of 34.5% and a total elongation of 40.6%. The nano-grained material would normally have a yield strength of up to 1200MPa, but with an elongation-to-failure of only 8.9%. It was concluded that the formation of a bimodal microstructure was closely related to the original ultrafine-grained or nanograined structure, with the dislocation density, stored energy and the degree of grain refinement being controlled by the severe plastic deformation. It was however not always possible to obtain a bimodal structure via annealing.

The strength-ductility paradox was evaded in nanostructured iron-based alloys by making copper and manganese additions[72]. A threefold (20%) ductility increase was obtained, together with a strength improvement of 100MPa over that of alloys with only copper additions. The strength-ductility improvement was attributed to interactions between transformation-induced plasticity, coherent nanoscale copper precipitates and heterogeneous stress-strain partitioning plus dislocation reactions.

The effect of copper additions to a Fe-Ni-Mo-Co-Cr maraging stainless steel was investigated[73] with regard to the aging behavior and its influence upon the mechanical properties. Copper-rich and molybdenum-rich phases precipitated sequentially from the steel matrix during aging. That is, a molybdenum-rich phase nucleated at a copper-rich phase and then grew. Together with segregation of copper and nickel, reverted austenite gradually formed. With increasing aging time, the stability of the reverted austenite increased, with a marked increase in its toughness. Following aging (90h), the yield and tensile strengths attained 1270 and 1495MPa, respectively. The impact energy and the fracture toughness were 81 and 102MPam$^{1/2}$, respectively.

Deformation-twin nucleation at the predominantly $\Sigma3\{111\}$ grain boundaries in an ultrafine-grained (0.79µm) carbon-free high-manganese twinning-induced plasticity steel, Fe-31Mn-3Al-3wt%Si, was studied[74]. During tensile stressing, grain-boundary sliding changed the structure of the coherent $\Sigma3\{111\}$ twin boundary from atomically smooth to partially defective. The formation of so-called disconnections on the $\Sigma3\{111\}$ boundaries was associated with the motion of Shockley partial dislocations. The disconnections acted as preferential nucleation sites for the deformation twin that is a characteristic departure from the coarse-grained equivalent. The main observation was that the $\Sigma3\{111\}$ twin boundaries could be nucleation sites for deformation twins, via a 3-step process which involved the formation of disconnections due to twin-boundary dislocation dynamics under an applied stress, followed by the multiplication and propagation of disconnections on the initially atomically-smooth boundaries and, finally, by an increasing localized strain at disconnections that provoked preferential deformation twin nucleation. A grain boundary could be an effective dislocation source if its rotation axis was well-aligned with the dislocation lines of Shockley partials or perfect dislocations such as with a [112] Σ21 tilt boundary or a [011] Σ9 tilt boundary, or if its excess free volume within grain-boundary regions could aid the formation of Shockley partial dislocations. There was an absence of grain-boundary dislocations and free volume in a coherent Σ3{111} twin boundary. The coherent Σ3{111} twin boundary was thus unlikely to act as an effective dislocation source. The coherent Σ3{111} twin boundaries therefore needed to be inherently defective or to become defective during plastic deformation. The structure of the coherent Σ3{111} twin boundary before tensile strain was perfectly atomically smooth, with no defective step (disconnection). The deformation twin nucleation mechanism on Σ3{111} twin boundaries in the ultrafine-grained steel was different to that in the coarse-grained equivalent. Deformation-twin nucleation could occur on all types of boundary, regardless of the grain-boundary misorientation. The high-angle grain boundaries having a particular tilt axis, and low-energy coherent twin boundaries could all be nucleation sites for deformation twins. The disconnections accumulated strain and were nucleation sites for deformation twins when the localized stress exceeded the twinning stress. The difference, between ultrafine-grained and coarse-grained alloys, in the generation of localized strain concentrations was deemed to originate from grain-size constraints which suppressed in-grain dislocation activity in the former material and stimulated twin boundary dislocation activity, leading to the twin boundary disconnections which then acted as deformation-twin nucleation sites. The relevance of the results was that the sequential activation of various deformation modes could encourage the regeneration of a strain-hardening capability during plastic deformation and thence lead to a high strength plus high ductility due to interactions between differing deformation modes.

The mechanical performance of low-carbon advanced nanostructured steels depends greatly upon microstructural features such as the size, type, number-density and distribution of precipitates. Such steels can possess a tensile strength of up to 2GPa, together with an elongation of more than 10% and a reduction-of-area of up to 40%. Nanostructured steels with good strengths and low-temperature impact toughness values can moreover be obtained[75] by avoiding the temper-embrittlement regime. By adding an optimum amount of manganese, the steels can be made to exhibit an increased ductility (about 30%) while maintaining a yield strength of 1100 to 1300MPa and an ultimate tensile strength of 1300 to 1400MPa. They also offered a good weldability. With regard to the strength-ductility paradox (figure 9), it was noted that dislocation-looping around non-deformable particles was more favourable for improved ductility because the slip-distribution was more fine and homogeneous. A dispersion of ultra-fine precipitates could restrict dislocation motion so as to form a stress-concentration and increase the stress required for crack nucleation. Brittleness could be a result of easy crack-nucleation due to large primary precipitates, with the cracks then propagating easily through fine dispersions of secondary precipitates without loss of energy. An ultra-fine dispersion of high number-density coherent precipitates could markedly improve the strength while hardly impairing the ductility. A ductility increase could also be attributed to a stress-homogenization which arose to the ultra-fine dispersion of high-number density coherent precipitates. A high coherency between precipitate and matrix reduced strain accumulation and prevented crack initiation as the dislocations sheared through the precipitates. A fine dispersion of nanoscale coherent precipitates could promote the simultaneous rotation of multiple crystal planes during deformation and lead to better stress-transfer, and thus to increased ductility.

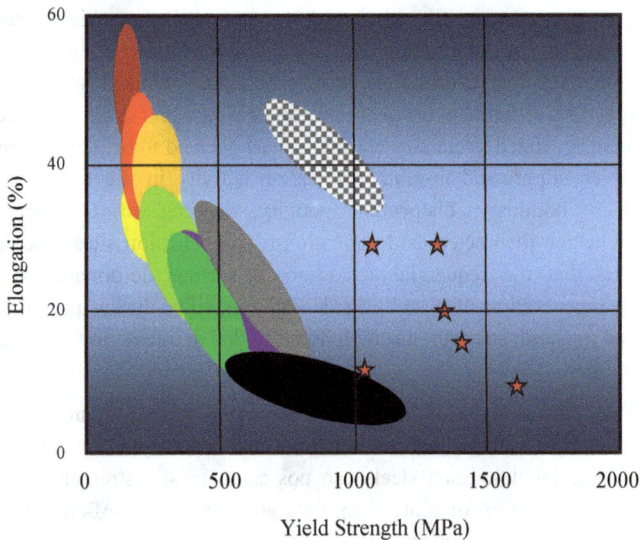

Figure 9. Siting new low-carbon advanced nanostructured steels among competing conventional commercial steels. Dark red: IF, light red: mild steel, orange: IF-HS, yellow: BH, light green: CM, dark green: high-speed low-alloy steels, purple: DC-CP, gray: transformation-induced plasticity steels, black: high-carbon martensitic steels, chequer: transformation-induced plasticity steels. Stars: low-carbon advanced nanostructured steels

Materials Research Forum LLC
https://doi.org/10.21741/9781644903230

When samples of AISI1045 0.45%C steel were subjected to 4 to 6 passes of equal-channel angular pressing at 400C (table 17), dynamic recovery occurred in the ferrite, with the formation of sub-grains which were some 320nm in size, plus isolated grains of sub-micron size with high-angle boundaries[76]. The size of the initial ferrite grains and pearlite colonies before deformation had been about 15μm. Fragmentation and partial spheroidization of cementite lamellae occurred in the pearlite colonies. Annealing (550C, 5h) led to essentially complete spheroidization of carbides, with an average size of about 280nm, and slightly increased the average ferrite grain size to about 410nm. The sub-micron grain structure led to a marked strengthening of the steel, to a yield strength of 960MPa, while maintaining an adequate elongation of 8%. Further annealing (550C, 5h) increased the ductility to 14% but decreased the yield strength to 745MPa. Discontinuous and wavy cementite-lamellae morphologies were present in the deformed structure, and the onset of lamellar fragmentation was visible after 4 passes. The onset of spheroidization was apparent after 6 passes. Also following 4 passes there was an imperfect sub-grain structure, with quite thick sub-grain boundaries, high densities of lattice and grain-boundary dislocations and an average sub-grain size of about 0.4μm. When the number of passes was increased to 6, the average sub-grain size decreased to about 0.3μm, while dynamic recovery led to sub-grain boundary-thinning and a decrease in the density of free dislocations. The localized appearance of high-angle boundaries indicated the onset of a transformation of the sub-grain structure into a sub-microcrystalline structure. Overall, 6 pressing passes increased the yield and ultimate tensile strengths by a factor of some 2.5 (to 960MPa) and 1.5 (to greater than 1GPa), respectively. The elongation decreased by more than a factor of 2 (to some 8%) and the yield ratio was then equal to 0.95. Annealing (500C, 3h) following pressing hardly changed the structure and the carbide spheroidization was very modest. Even 50 hours of annealing failed to complete spheroidization, and the carbide distribution remained non-uniform. Annealing at 500C after 6, rather than 4, passes, led to more notable cementite-lamellae fragmentation while spheroidization within and without pearlite colonies occurred more rapidly. The carbides outside of pearlite colonies, after annealing for 50h, were 0.4μm in size rather than 0.3μm. The average carbide size within pearlite colonies, following annealing for 50h, was 0.15μm for both states.

Table 17. Strength and ductility of 0.45%C steel following equal-angle angular pressing

Passes	Annealing	YS(MPa)	UTS(MPa)	Elongation(%)	Reduction-of-Area(%)
0	normalization	402	725	18.1	22
4	-	892	981	6.3	43
6	-	960	1013	7.7	31
6	500C, 10h	798	829	12.6	51
6	550C, 5h	746	830	13.8	45

A new 2-step heat-treatment, termed short annealing and tempering, was devised[77] which involved short-term (120s) annealing and tempering, and was used to investigate the interactive effect of austenite reversion and nano-precipitation upon the tensile properties of very-low carbon steel containing 1.5wt% of copper and 1.5wt% of nickel. Annealing produced 33vol% of reverted austenite, and the time was minimised in order to prevent the copper and nickel from partitioning into austenite, but rather remain in the ferrite and promote nano-precipitation during the following tempering. During the latter step, nano-precipitates with a copper concentration of 20 to 50at% in the precipitation core and enriched with copper, nickel, aluminium and manganese were observed in the ferrite. The volume fraction of reverted austenite attained 38.5vol% during tempering and imparted an ultimate tensile strength of 1222MPa, combined with a total elongation of 29%. The large amount of reverted austenite, plus the nano-precipitates, improved the strain-hardening behavior as well as the tensile properties. The increase in yield strength was attributed mainly to the nano-precipitates, while the increase in ultimate tensile strength and total elongation were improved by the onset of transformation-induced plasticity; as controlled by the stability of the modified austenite. The volume fraction and mechanical stability of the austenite were key factors in controlling the tensile strength and elongation. A reduction-of-area mechanical stability coefficient, k, was defined in order to express the mechanical stability of the reduction-of-area in terms of its value following necking: $f_a = f_b e^{-k\varepsilon}$, where f_a and f_b were the austenite fraction after and before deformation to a strain of ε, respectively. A high k-value tended to correspond to a low mechanical stability of the reduction-of-area (table 18). The SAT specimens had undergone the 2-step process with no intermediate quenching before being annealed (700C, 120s) and tempered (500C, 3h) before water-quenching to room temperature. The SA specimens were annealed (700C, 120s) and water-quenched to ambient. The tensile

Materials Research Forum LLC

https://doi.org/10.21741/9781644903230

properties (table 19) of one sample were a yield strength of 987MPa and an ultimate tensile strength of 1257MPa, but the ductility was only 4.1% (uniform elongation) and 12.7% (total elongation). Following the SA treatment, the 1.5Cu-1.5Ni SA sample had a yield strength of only 577MPa and an ultimate tensile strength of 1107MPa. The ductility of the SA specimen was improved, with a uniform elongation of 15.5% and total elongation of 24.9%. Following the SAT treatment, the 1.5Cu-1.5Ni specimen had a yield strength of 848MPa and an ultimate tensile strength of 1222MPa. The ductility of the 1.5Cu-1.5Ni SAT specimen was such that it had a uniform elongation of 20.1% and a total elongation of 29.1%. Unlike the 1.5Cu-1.5Ni SA specimen, the 1.5Cu-1.5Ni SAT specimen offered a remarkable combination of a greatly increased strength and an improved ductility. The strain-hardening of all of the specimens benefited from an appreciable transformation-induced plasticity effect. The strain-hardening behaviour of the 1.5Cu-1.5Ni SAT sample also benefited from the interaction of dislocations and nano-precipitates.

Table 18. Mechanical stability coefficient of low-carbon copper-nickel steels

Sample	k
0Cu-0NiSA	12.4
1.5Cu-1.5NiSAT	12.3
1.5Cu-1.5NiSA	9.0
0Cu-0NiSAT	30.9
0Cu-0NiSA	644

Table 19. Tensile properties of low-carbon copper-nickel steels

Sample	YS(MPa)	UTS(MPa)	$e_{uniform}$(%)	e_{total}(%)
1.5Cu-1.5NiAF	987	1257	4	13
0Cu-0NiAF	849	1123	4	12
1.5Cu-1.5NiSAT	848	1222	21	29
1.5Cu-1.5NiSA	577	1107	16	25
0Cu-0NiSAT	692	996	18	28
0Cu-0NiSA	644	964	15	23

AF: as-forged

Close examination of the ductility of sheet metal has revealed a further paradox. Specimens can exhibit localized necking or fracture; termed global and local formability, respectively. Forming limit curve and uniaxial tensile tests determine the global formability while fracture forming limits, true-thickness fracture strains and other criteria measure the local formability. Dual-phase steels of similar strength tend however to be ductile in tensile tests but exhibit a low ductility in hole-expansion tests. On the other hand, other steels exhibit a low ductility in conventional tests but much better results in other tests. A statistical analysis was made[78] of model 2-phase microstructures by scaling the hardness of martensite and its volume fraction in order to generate model dual-phase steels possessing the same strength but exhibiting differing strain-hardening and hardness-differences. The model reproduced the experimentally observed behaviours. In dual-phase steels, a greater global (necking-related) ductility was found after delaying martensite plasticity by increasing the hardness-difference of the two phases. The stress-strain distributions then became more heterogeneous and led to a lower fracture limit of one of the phases or interfaces under shear loading, and reduced the local ductility. The global ductility was improved by a greater hardness-difference and a lower martensite volume fraction, but the local ductility was improved by a small hardness-difference and a higher martensite volume fraction. The hardening behaviour of the martensite was the critical factor in avoiding the compromise between local and global ductilities. If the strain-hardening capability of the martensite at high strains could be increased, it would evade the observed paradox of local versus global ductility in dual-phase steels.

Gradient-structured 301 stainless steel was produced[79] via surface mechanical attrition, using high-speed 3mm steel shot. The gradient microstructure ranged from a nanostructured surface with a high volume fraction of martensite, to a coarse-grained austenite interior. There was high hardness gradient across all of the gradient samples, indicating an appreciable mechanical incompatibility between the surface layers and the central matrix. The mechanical gradient became more marked as the processing time increased. After 600s, the hardness ranged from 560HV at the surface to 380HV at the centre. The material exhibited a very good strength-ductility combination due to a gradient-induced hardening and a phase-transformation induced plasticity. There was a continuous $\gamma \rightarrow \alpha'$-martensite transformation in the surface layers during tensile straining. Differences in strength and strain-hardening ability between neighbouring layers can introduce a strain inhomogeneity between them. This could then produce a constraint between layers and a gradient strain-distribution across the layer boundary. An accumulation of geometrically necessary dislocations in the strain-gradient zone led to the development of long-range internal stresses which promoted both strengthening and strain-hardening. Transformation-induced plasticity is itself an effective means for improving strain-hardening: it suppresses the development of plastic damage by delaying strain-localization and introduces a strain-gradient at phase boundaries which maintains strain-continuity. Geometrically necessary dislocations then accumulate around the phase boundary and accommodate the plastic-strain gradient; further improving strength and strain-hardening.

Depth-dependent microstructures were studied[80] in fully austenitic gradient nanostructured 310S stainless steel which had been prepared by means of surface mechanical rolling. This was done in such a way that the severe plastic deformation did not provoke martensitic transformation. The gradient nanostructured surface layer which formed at room temperature had a grain size that ranged from about 56nm to tens of microns in going from surface to the core. The deformed surface layer increased in thickness with increasing numbers of treatment passes and pressure, with the final thickness being some 1250µm. The stress in the surface layer after 9 passes was compressive, while the core stress was tensile. As the tensile loading was increased, the core material first plastically yielded and strain-hardened, and the yield area expanded from the centre to the surface until the nanograins on the outermost layer underwent plastic yielding. The hardness of the 9-pass material exhibited a decrease along the radius of the cross-section. The surface layer had little effect upon the strain-hardening of the core. The surface layer consisted of a microstructurally refined region of less than some 500µm and a work-hardened region of 500 to 1250µm. The deformation was dominated by the formation and evolution of dislocation structures, twins and shear-bands. When compared with coarse-grained material, the hardness of the uppermost surface was increased two-fold and the yield and ultimate strengths were increased by 100% and

18.72%, respectively. The gradient distribution of residual stress fields, work-hardening properties and material strength governed the strength-ductility interactions.

Materials having a gradient microstructure can also exhibit a good combination of strength and ductility. A heterostructured 304 stainless steel which had a gradient dislocation structure within micron-sized grains, that was produced by cyclic-torsion processing, possessed a much improved yield strength and a slightly-reduced uniform elongation, as compared with those of a coarse-grained equivalent[81]. The grain orientations, from the surface to the core, remained as random as they were before cyclic torsion, with no obvious grain refinement. Dense high-angle grain boundaries were found following the treatment. Many low-angle boundaries with misorientations of less than 15° were introduced into the grain interiors, and their density decreased with increasing depth from the top surface. The gradient low-angle boundaries within unchanged coarse grains were very different to the ultra-fine and nano-sized grains found in the usual homogeneous or gradient nanostructures. Different dislocation patterns were introduced into the grain interiors. There were many dislocation cells at the uppermost (~0.02mm) surface, with a high density of dislocations at the cell-walls and fewer dislocations in the interior. With increasing depth (~0.5mm), the dislocation pattern gradually changed to the planar single-slip induced dislocation walls pattern which is common in face-centred cubic metals, with a low stacking-fault energy deformed to large strains. The 0.2% yield strength was 464MPa; some 2.2 times that (210MPa) of the dislocation-free coarse-grained equivalent, with a uniform elongation of up to 55%. This was slightly poorer than that of the equivalent. The strain-hardening ability was slightly higher than that of the dislocation-free coarse-grained material at the same stress. Due to the macroscopic dislocation gradient structure, as-prepared samples had a HV distribution which ranged from 3.6GPa at the uppermost surface to 1.9GPa at the core. The net value for dislocation-free coarse-grained material was about 1.5GPa. Tensile straining led to a monotonic hardening of the gradient material, from the uppermost surface to the core. The microhardness, in going from the uppermost surface layer to the core of the gradient material, ranged from 4.3 to 3.6GPa at a strain of 40%. The density of geometrically necessary dislocations at low-angle boundaries in the uppermost layer was about $4.0 \times 10^{14}/m^2$. This was slightly higher than that before tension. The improved mechanical properties were attributed mainly to the gradient-induced strain delocalization, and the associated activation of various strengthening mechanisms, such as extensive dislocation interaction and the formation of dense dislocation patterns, nanotwins and martensitic phases.

The effects of nickel and aluminium additions upon the strength-ductility relationship of copper-containing high-manganese steels were investigated[82]. The strength and total elongation of the original steel were both markedly increased by the additions and by aging. The strength-ductility combination of the modified steels was best following aging (2h), and this time was naturally used for all of the studies. The data (table 20) indicated an increase in both strength and total elongation due to the Ni/Al additions. The solid-solution treated NiAl steel also exhibited delayed serrated flow. Aged NiAl samples had a higher yield strength, tensile strength and elongation than those of solid-solution treated samples. In the original steel, serrated flow of aged samples was delayed as compared to solid-solution treated material. In NiAl steel, serrated flow of the aged material occurred at lower strains, suggesting differing effects of Ni/Al additions and aging upon serrated flow. This was attributed to phase transformation and the formation of precipitates. The fracture surfaces of aged NiAl-steel had larger, uniform and more continuous dimples than those of the original steel, thus indicating good plasticity. The strain-hardening rates of solid-solution treated samples were lower than those of aged samples. In the pre-tensioned original steel, plate-like α-martensite was surrounded by a large amount of austenite (table 21). In the post-tension original steel the martensite fraction had markedly increased and large amounts of needle-like ε-martensite were introduced by the uniaxial tension. Such changes were common in the solid-solution treated original and NiAl steels, and led to transformation-induced plasticity. In post-tension NiAl steel the martensite fraction increased, and deformation twins were introduced, leading to twinning-induced plasticity. In NiAl steel, the volume-fraction of martensite phases steadily increased and deformation-twins formed with increasing external tensile stress (table 22). This greatly retarded necking, and thus improved strength and ductility. In NiAl steel, dislocations were arrayed on deformation-twin boundaries and piled-up at the boundaries. The dislocations slipped on the twin-boundary planes and contributed to plastic strain. The yield strength of the high-manganese steels was greatly increased by aging. Similar results were found for the original steel. An increase in yield strength was attributed to nano-precipitation strengthening, phase transformation and grain-refinement during aging. Copper-rich precipitates appeared in aged Cu-bearing high-manganese steel and greatly affected the strength. Overall, the improvement in the strength-ductility combination due to additions was much greater than that due to aging.

Materials Research Forum LLC
https://doi.org/10.21741/9781644903230

Table 20. Room-temperature properties of high-manganese steels

Steel	Aging Time (min)	YS(MPa)	UTS(MPa)	e(%)
original	0	254	611	25
original	120	290	768	31
NiAl-added	0	243	716	73
NiAl-added	120	275	771	77

Table 21. Variation in volume-fraction of austenite with strain

Steel	Condition	Strain(%)	Austenite(vol%)
original	solid-solution	0	78
original	aged	0	97
NiAl-added	solid-solution	0	100
NiAl-added	aged	0	98
original	solid-solution	0.18	45
original	aged	0.18	56
NiAl-added	solid-solution	0.18	79
original	solid-solution	0.25	36
original	aged	0.31	42
NiAl-added	aged	0.40	73
NiAl-added	solid-solution	0.73	70
NiAl-added	aged	0.77	57

Table 22. Fractions of twins in steels

Steel	Condition	Twins(%)
original	solid-solution	47.0
original	aged	42.7
NiAl-added	solid-solution	45.8
NiAl-added	aged	42.7

High-Entropy Alloys

A review of this alloy group emphasised[83] that heterogeneous structures (table 23) could increase the strain-hardening capacity, and delay plastic instability, thus evading the paradox. After surveying the studied high-entropy and medium-entropy alloys, it was noted that they should comprise various microstructures having large strength differences. This then led to stress- and strain-partitioning during tensile deformation, and thence to pile-ups of geometrically necessary dislocations and hetero-deformation induced stress effects. An important strategy is the use of gradient structures (table 24). A microstructural gradient at the macroscopic scale involves a gradually increasing grain size, increasing sub-structure size and a reduced defect-density in going from the surface to the interior. The gradient structures can exhibit a marked mechanical incompatibility which then leads to a macroscopic strain-gradient and stress state which is resolved by accumulations of geometrically necessary dislocations. Such gradient structures have good strength and ductility combinations. The gradients are commonly created by the use of surface mechanical grinding, surface mechanical rolling, surface mechanical attrition and rotationally accelerated shot-peening. These treatments produce nanostructured surface layers having a depth of only a few hundred μm. The use of high-pressure torsion can produce large-scale gradients across the diameter of a specimen. Increased strength and ductility can be obtained by using a gradient structure with an undeformed core between thin deformed surface layers. Asymmetrical rolling, followed by annealing, is another means for producing gradient microstructures.

Table 23. Heterogeneous structures and properties of medium- and high-entropy alloys

Alloy	Structure	Test	YS (MPa)	UTS (MPa)	e_u (%)
$Co_{25}Ni_{25}Fe_{25}Al_{7.5}Cu_{17.5}$	NG, UFG	compressive	1795	1936	10.6
$Ti_{10}Fe_{30}Co_{30}Ni_{30}$	NG, CG	compressive	1830	2024	18.7
$Cr_{20}Fe_6Co_{34}Ni_{34}Mo_6$	FG, UFG	tensile	1100	1300	29
$V_{10}Cr_{15}Mn_5Fe_{35}Co_{10}Ni_{25}$	FG, CG	tensile	761	936	28.3
$Al_{0.1}CoCrFeNi$	UFG, FG, CG	tensile	711	928	30.3
$Al_{0.1}CoCrFeNi$	FG, CG	tensile	525	784	37
CrMnFeCoNi	FG, UFG	tensile	625	855	50.7
CoCrFeNiMn	UFG, CG	tensile	1298	1390	9.4
CoCrNi	NG, UFG, FG	tensile	1150	1320	22
CoCrNi	UFG, FG	tensile	797	1360	19
CoCrNi	FG, UFG	tensile	928	1191	28
CrCoNi	CG, NG	tensile	1452	1520	10
CoCrNi	UFG, FG	tensile	1435	1580	24

CG: coarse grains, FG: fine grains, UFG: ultra-fine grains, NG: nanograins

It was noted[84] that, although the strength-ductility compromise represented by the blue curve (figure 10) was an improvement on that for conventional alloys, it nevertheless left – literally – much room for improvement; as indicated by the red curve. The data points defining the blue curve represent alloys which all exhibit a large degree of nanoscale heterogeneity, thus suggesting that a deliberate increase in heterogeneity might promote inhomogeneous plastic deformation and move the strength–ductility curve towards the right-hand side. Precipitation-hardened alloys, multiphase alloys and composites can already exhibit high strength plus good ductility, and they are heterogeneous. On the other hand, the remaining scope for employing familiar mechanisms is limited when the alloy consists of a single major element. It was this fact which drew attention to the high-entropy alloys; also known as multi principal component alloys and complex concentrated alloys.

Table 24. Properties of medium- and high-entropy alloys with a gradient structure

Alloy	Treatment	YS (MPa)	UTS (MPa)	e_u (%)
CoCrFeMnNi	ASR, ann	700	930	42
$Al_{0.1}CoCrFeNi$	CDT	510	850	19
$Al_{0.1}CoCrFeNi$	CT	539	690	42
$(Fe_{40}Mn_{40}Co_{10}Cr_{10})_{96.7}C_{3.3}$	SMRT	587	885	40.4
$(Fe_{40}Mn_{40}Co_{10}Cr_{10})_{96.7}C_{3.3}$	SMRT	765	956	20.5
CrCoFeNiMn	RASP	610	680	15
CoCrFeMnNi	EP-USR	750	802	21.9
$CoCrFeNiMo_{0.15}$	Torsion	724	904	27
$FeCoCrNiMo_{0.15}$	SP	486	855	46.8
CoCrNi	Torsion	760	880	31
CoCrNi	Torsion, ann	930	1050	27

ASR: asymmetrical rolling, ann: annealing, EP-USR: electropulse-assisted ultrasonic surface rolling, SP: shot-peening. CDT: cyclic dynamic torsion, SMRT: surface mechanical rolling technique, RASP: rotationally accelerated shot-peening

High-entropy alloys were discovered only some two decades ago and are now hoped to constitute another material group which can evade the paradox. Not being constrained by the venerable Hume-Rothery rules, they offer a wider choice of compositions and thus the increased probability of finding new alloys possessing novel properties. The high-entropy alloys comprise many elements in almost equal concentrations, rather than a single element plus increasingly smaller additions of other elements. As usual, these strong alloys tend to suffer from limited ductility and poor work-hardenability, although strain-induced deformation mechanisms such as twinning-induced and transformation-induced plasticity can improve the poor work-hardenability at the cost of often lower yield-strengths.

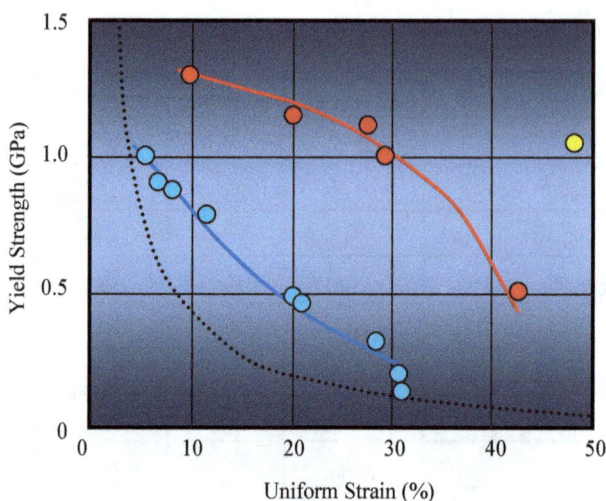

Figure 10. Yield strength versus uniform tensile strain. The blue curve shows data for heterogeneous nanostructured materials and the red curve shows data for heterogeneous high-entropy alloys. The isolated green dot[1] represents data for a complex alloy comprising multi-component intermetallic nanoparticles. Everything to the left of, or below the dotted curve is a simple metal

Multi-level heterogeneities are possible in high-entropy alloys. The first level of heterogeneity arises from the several constituents which are present in a concentrated solution at the atomic-level. There can also be an appreciable inhomogeneity which fluctuates from place to place, and variations in composition and packing of the various elements. Local chemical order can develop at the short or medium range. The next level of heterogeneity involves minute closely-spaced clusters and complexes, plus precursors of precipitates, rather like Guinier-Preston zones that may appear at the nanometre scale before a second phase becomes obvious in a previously single-phase solution. The third level comprises multiphase nanostructures that evolve from the parent solution, thus rendering part of the alloy dual-phased. These may be precipitates, lamellar eutectic structures or martensites. The fourth level arises from defects which are embedded within the lattice, and include nanotwins, stacking-faults and grain boundaries. At a final level,

---

[1] It consisted of a high (up to 55%) volume fraction of ductile nanoparticles which were coherent with, and uniformly distributed within, the face-centred cubic FeCoNi matrix. They had an $L1_2$ crystal structure and the composition, $(Ni_{43.3}Co_{23.7}Fe_8)_3(Ti_{14.4}Al_{8.6}Fe_2)$.

the grain-size distribution can be deliberately made multimodal or rendered gradient-like, with length-scales ranging from a few nanometres to micrometres. These 5 levels of heterogeneity interact in a given high-entropy alloy and tend to make them unusually vulnerable to non-uniform plastic deformation even when the applied stress is uniform. The alloy represented by the green spot in the above figure attained an elongation of about 50%, combined with a yield stress of more than 1GPa; a striking example of the properties that high-entropy alloys can offer.

In face-centred cubic high-entropy alloys there persists a compromise between yield strength and elongation. The combinations are nevertheless better than those of single-phase heterogeneous nanoscale materials. The metastable nature of the high-entropy alloys can lead to phase decomposition and the production of intermetallic hardening precipitates, plus a mechanically-promoted transformation to give multiphase microstructures. Some alloys contain heterogeneities which arise from precipitation hardening. In this case the precipitates interact with dislocations and impart both strength and an increased ability to strain-harden. The hard and brittle intermetallic particles do not lead to fracture because the toughening effect of the surrounding matrix and its excellent ductility. Another multiphase strategy involves dual-phase eutectic lamellae. Yet another strategy is the deliberate lowering of the stacking-fault energy by adjusting the composition. The best property combination again appears to be that corresponding to the green point in the above figure. The properties were attributed to multistage work-hardening, with a marked tendency of the dislocations to interact with the uniformly distributed high-density interstitial precipitates. In addition to second-phase heterogeneities, there can also be heterogeneities which are introduced by deformation. Directional dislocation sub-structures, such as dense dislocation arrays and walls, create long-range back-stresses. These then contribute to strain-hardening, in addition to the large increase in short-range stresses arising from accumulated forest dislocations. Micro-bands which resemble low-angle grain boundaries are also introduced by deformation. With regard to the strength–ductility behaviour of body-centred cubic high-entropy alloys, a lot of interest has focused upon single-phase solid solutions involving refractory metals such as titanium, zirconium, hafnium, vanadium, niobium, tantalum, molybdenum and tungsten. At room temperature, the refractory alloys exhibit yield strengths of up to 2GPa, and most of them have some ductility only in compression. The lack of room-temperature ductility is related to a sharp increase in the ductile-to-brittle transition temperature, where the fracture stress and the yield stress cross, when large fractions of alloying elements are added to a first body-centred cubic initial element. Those refractory alloys which exhibit some tensile ductility are based mainly upon the Ti–Zr–Hf–Nb–Ta system and include equi-atomic TiZrHfNbTa with its yield strength of ~830MPa and tensile elongation-to-failure of some 9%. Adding heterogeneities to body-centred cubic high-entropy alloys can be an effective means for simultaneously increasing strength and strain-hardening. Precipitates have rarely been used in these materials. On the other hand, nanoscale clusters have been introduced by adding ~2at% of oxygen to TiZrHfNb. It interacted preferentially with titanium and zirconium and segregated so as to form a high density of nanoscale (O, Ti, Zr) complexes of a few nanometres in size and a few nanometres apart. The complexes then interacted with mobile dislocations, thereby increasing the strength while increasing strain-hardening and ductility. This was attributed to the pinning of the dislocations by the oxygen complexes, and to the homogenization of strains via the wavy slip of dislocations. This aided cross-slip and dislocation multiplication, rather than the alternative of planar slip which would have localized strain. A ductility increase to ~28%, due to oxygen complexes, merely served however to restore the tensile elongation of the non-strengthened metal. The ductility was also not due to uniform straining, but included very diffuse necking; a known feature of high-strength body-centred cubic alloys.

Another possibility is to strengthen high-entropy alloys by means of short-range and long-range ordering, while preserving good strain-hardenability and ductility. An understanding of local chemical ordering in random solid solutions is required in order to guide the design of these multi-component alloys. A simple thermodynamic model which is based entirely upon the binary mixing enthalpies of mixing can select the optimum alloy components for controlling the extent of chemical ordering in high-entropy alloys.

In one study[85], high-resolution electron microscopy, atom-probe tomography, Monte Carlo methods, quasi-random structures and density functional theory were combined in order to show how additions of aluminium and titanium, and suitable annealing, affects the chemical ordering in nearly-random equi-atomic face-centred cubic CoFeNi solid solutions. Short-range ordered domains, which were the precursors of long-range ordered precipitates, determined the mechanical properties. In particular, an increase in the local order increased the tensile yield strength of the original CoFeNi alloy by a factor of 4 while also greatly improving the ductility (table 25), thus evading the paradox.

Table 25. Effect of ordering upon the tensile properties of high-entropy alloys

Alloy	Ordering	Yield Stress (MPa)	UTS (MPa)	Elongation (%)
CoFeNi	none	181	482	21
$Al_{0.25}CoFeNi$	short-range	287	582	47
$Al_{0.3}CoFeNi$	early long-range	454	853	76
$Al_{0.3}Ti_{0.2}CoFeNi$	long-range	802	1121	34

The almost ideal CoFeNi random face-centred cubic solid solution thus had a basic yield strength of 181MPa while the $Al_{0.25}CoFeNi$ alloy, with its short-range order exhibited the higher yield strength of 287MPa. The introduction of ~3nm ordered domains (early-stage long-range order) into the $Al_{0.3}CoFeNi$ alloy increased the yield strength to 454MPa. The introduction of well-developed domains of long-range order into $Al_{0.3}Ti_{0.2}CoFeNi$ increased the yield strength to 802MPa. The yield and tensile strengths thus increased with increasing chemical order. The ductility meanwhile increased from ~21% for CoFeNi to ~76% for $Al_{0.3}CoFeNi$, thus showing that increasing short-range order, or the early stages of long-range order increased the ductility over that of the initial near-random solid solution. It was further noted that the work-hardening rate of the equi-atomic CoFeNi system constantly decreased until the failure strain was reached; the work-hardening behaviour exhibited by single-phase face-centred cubic alloys. The high-entropy alloys which exhibited chemical ordering also exhibited a marked change in the work-hardening slope; reflecting an increase in ductility. The size-mismatch of aluminium and titanium with respect to the cobalt, iron and nickel atoms was expected to produce solid-solution strengthening, but the calculated solid-solution strengthening effect was only 20 to 30MPa; less than the increases due to ordering. Post-deformation microscopy of CoFeNi which was deformed to failure revealed no sign of deformation-induced twinning or phase transformation during tensile testing. The deformation of CoFeNi occurred via homogeneous dislocation-mediated plasticity. Planar arrays of dislocations were seen during the early stages of plastic strain in $Al_{0.25}CoFeNi$ and it exhibited homogeneous plasticity when deformed to failure. The deformation of the ordered alloys appeared to begin with a glide-plane softening which localized the slip into {111} planes. Subsequent interaction between the localized slip-bands then refined the slip-length and led to a dynamic Hall–Petch-like type of effect. Homogeneous dislocation–mediated plasticity began when a critical strain was attained within the planar slip-bands. These effects corresponded to the improvements in strain-hardening and elongation. It was also demonstrated that additions of aluminium, which exhibited large negative enthalpies of mixing with respect to the constituents of the nearly-random body-centred cubic NbTaTi high-entropy alloy, again led to chemical ordering and to improved mechanical properties.

It was further pointed out[86] that the introduction of chemical ordering, the interaction between planar slip on non-parallel {111} planes and the dynamic Hall-Petch like effect which refined the slip length did not work in hexagonal close-packed alloys. Such glide-plane softening there tended to cause catastrophic failure. The strengthening ability of short-range and long-range ordered domains results from their acting as obstacles to glissile dislocations, due to the associated strain-fields. The shearing of long-range ordered domains by dislocations is also energetically untenable due to the formation of antiphase boundaries and planar faults such as super extrinsic and super intrinsic stacking faults. In traditional face-centred cubic alloys, such as nickel-based superalloys, the associated fault energies are very high; thus making precipitate-shearing difficult and leading to higher yield strengths. Short-range and long-range order are generally avoided in hexagonal close-packed alloys, such as those of titanium, because they lead to slip-planarity and poor strain-hardening; also known as glide-plane softening. The onset of planar slip tends to be less detrimental in dilute face-centred cubic alloys.

Samples of cobalt-rich $Co_{35}Cr_{32}Ni_{27}$-Al_3Ti_3 high-entropy alloy which were prepared by hot-forging and annealing (700C, 8h) had a microstructure which comprised phases having a face-centred cubic, an hexagonal close-packed and a $L1_2$ structure[87]. The CALPHAD method had been used to develop the alloy. Low densities of $L1_2$ particles with an average diameter of 15nm precipitated in the matrix of what was essentially a dual-phase microstructure of cubic and hexagonal phases; both of which had compositions that were similar to $Co_{39}Cr_{35}Ni_{22}Al_2Ti_2$. The $L1_2$ precipitates were $Ni_{60}Co_{10}Cr_5Al_{10}Ti_{15}$. The nano-lamellar or small-block hexagonal phase was dispersed within the cubic phase at low volume fractions. The mechanical properties of the 3-phase alloy at 293K featured a yield stress of the order of 1.12GPa and an ultimate tensile stress of some 1.4GPa, while the tensile elongation was 36%. At 77K, the strength-ductility combination was better, with a yield strength, ultimate tensile strength and elongation of 1.3GPa, 1.8GPa and 53%, respectively. The strain-hardening rates of aging alloys at 293 and 77K exhibited a non-monotonic evolution. At 77K, the alloy exhibited a steady hardening rate of some 4800MPa up to a true strain of 0.2. The strain-hardening rate then gradually decreased to about 3000MPa at a true strain of 0.40. It was emphasized that a high yield stress and high elongation existed at both ambient and cryogenic temperatures. The high yield stress was attributed to solid-solution strengthening, grain-refinement

Materials Research Forum LLC
https://doi.org/10.21741/9781644903230

strengthening, precipitation-strengthening and second-phase strengthening. The addition of 3at% of aluminium and titanium was estimated to contribute about 175MPa of solid-solution hardening. When accounting for the Hall-Petch effect, the yield stress of samples with a grain size of 10μm was estimated to comprise a lattice friction of 218MPa plus 568MPa/$\mu m^{-1/2}$ for the grain size. The overall contribution to the yield stress, arising from grain-size strengthening, was deduced to be about 390MPa. The precipitation-strengthening was ascribed largely to a marked order-strengthening of the γ' particles, leading to an estimated contribution of 310MPa. The total estimated contribution to the yield stress, arising from solid-solution strengthening, grain-refinement and precipitation-strengthening was therefore 875MPa. This was far lower than the measured room-temperature yield stress of 1210MPa. It was proposed that the hexagonal phase, dispersed within the cubic matrix, explained the additional strengthening. The yield stress at 77K was slightly higher than that at 293K, and this was attributed to a higher friction stress at 77K.

The low-cycle fatigue behaviour[88] of the metastable $Fe_{50}Mn_{30}Co_{10}Cr_{10}$ dual-phase (face-centred cubic and hexagonal close-packed) high-entropy alloy exhibited a deformation-induced martensitic transformation from the cubic to the hexagonal phase at strain amplitudes of 0.3% and 0.6%. At strain amplitudes of 0.9% and 1.2%, there was extensive twinning in the hexagonal phase, together with some hexagonal to cubic transformation. The occurrence of planar slip and deformation twinning in both phases, together with bidirectional phase transformation led to an hierarchically-refined microstructure. This permitted a certain tailoring of the structure in order to optimise both strength and ductility. It was possible to provide sufficient obstacles to dislocation motion in order to impart a high strength and yet also make mobile dislocations available at higher strains in order to impart sufficient ductility. Compositional tuning could reduce the stacking-fault energy and enable shear mechanisms like twinning and martensitic transformation in addition to dislocation slip. Tensile deformation of the alloy was characterized by a high strain-hardening due to lattice distortion, local concentration variations, solid-solution hardening, local stacking-fault energy variations, martensitic transformations and deformation twinning. Dynamic Hall-Petch strengthening led to good strength-ductility combinations.

The γ/γ' FeCoNiAlTi high-entropy alloys also evade the strength-ductility paradox and exhibit good combinations of strength and ductility. In order to clarify the origin of these results, calculations were made[89] of the generalized stacking-fault energies of the γ and γ' phases, of the relationship between planar stacking-faults and work-hardening and of the effect of chemical concentration and grain orientation upon deformation mechanisms. For both the face-centred cubic γ matrix and the $L1_2$ γ' precipitates, <110>{111} shear deformation was known to be the dominant slip-system and the Burgers vectors of complete dislocations in the matrix and L1$_2$ phases were a/2<110> and a<110>, respectively. In some cases, the complete dislocation might dissociate into 2 Shockley partial dislocations, with a stacking-fault connecting them. It was noted that the multicomponent nature of the material lowered the generalised stacking-fault energies of the matrix but increased those of the precipitates. The ratio of the intrinsic stacking-fault energy to the antiphase boundary energy was used to analyse the activation of micro-bands and planar stacking-faults in the γ/γ' alloys. It was found that a ratio of about 0.2 was optimum for imparting extended micro-band induced plasticity. Increasing the content of cobalt and titanium improved the strength-ductility balance, and facilitated micro-band activation by altering the stacking-fault energies of both γ and γ'. The micro-bands acted as impenetrable barriers which hindered dislocation glide and resulted in high strain-hardening. In the case of the γ matrix, it was assumed that all 5 of the elements were uniformly distributed on the same sub-lattice of the structure. In the case of the γ' precipitate, there were 2 different sub-lattices in the A$_3$B-type L1$_2$ structure, with nickel, cobalt and iron occupying the A-sites and aluminium and titanium occupying the B-sites. It was concluded that increasing the Co/Fe ratio or decreasing the Al/Ti ratio imparted a desirable combination of properties to this alloy.

Ultra-fine and nanocrystalline equi-atomic face-centred cubic high-entropy alloy, CoCrFeNiMn, had previously been produced[90] by means of high-pressure torsion at 300K or 77K. The hardness following room-temperature treatment was very high, whereas that following low-temperature treatment was distinctly lower, if the torsional strain was greater than 25. The values remained stable during long-term storage at room temperature. A similarly anomalous result was exhibited by *in situ* torque data, measured during high-pressure torsion. It was attributed to the increased hydrostatic pressure, to the low temperature and to the high shear-strains that were possible during low-temperature processing. Selected-area electron diffraction patterns indicated that a partial local change from face-centred cubic to hexagonal close-packed occurred under those conditions and led to a highly heterogeneous structure that exhibited an increase in the average grain-size and a marked decrease in the average dislocation density. This was deemed to be the cause of the anomalously low strength.

Gaseous carburization at 470C was used to surface-harden a CoCrFeNi high-entropy alloy[91], in that a carburized case of expanded face-centred cubic material, a carbide-free supersaturated interstitial solid solution of carbon (circa 3.0wt%), formed at the surface. Due to solid-solution strengthening by the interstitial carbon atoms and to strain-hardening caused by plastic accommodation of the lattice expansion, the surface hardness was markedly increased and attained 1221HV$_{0.1N}$; some 5 times harder than the substrate.

The alloy itself exhibited good ductility, even at cryogenic temperatures, but offered only a moderate surface hardness and yield strength. This improved the overall strength-ductility compromise.

A simple technique was proposed[92] for improving the strength-ductility compromise in the case of AlCoCrFeNi high-entropy alloy. The as-cast material had a coarse-grained microstructure comprising body-centred cubic and B2 phases. The alloy underwent spinodal decomposition into the AlNi-rich B2 phase and the CoCrFe-rich phase. The microstructure had an average grain size of about 92μm. The as-cast alloy was subjected to severe plastic deformation using stationary friction processing. Application of the process for just 0.25h led to an order-of-magnitude (45-fold) reduction in the grain size, and to a 53% phase transformation from body-centred to face-centred cubic. The stirred region of the processed specimen had a fine-grained microstructure which extended to a depth of some 800μm. The processed sample also exhibited thin elongated grains in the thermomechanically affected zone. The processed sample also exhibited a more-than-doubled ultimate tensile strength (650MPa) when compared to that (310MPa) of the as-cast alloy. Grain boundary strengthening, dislocation strengthening and precipitation strengthening were the main strengthening mechanisms. The ductility of the processed alloy was also increased from 11% to 18%. The combination of the fine grain-structure and the phase transformation was responsible for the exceptional mechanical properties. There seems to be no reason why this method should not be applied to other alloys so as to impart both high tensile strength and high ductility. Methods which increase strength also tend to undermine ductility, due to the hindrance of dislocation motion. The present case seems to be one of those in which a good strength-ductility compromise can be achieved via precise microstructural control involving doped heterogeneous structures, lamella microstructures or bi-modular microstructures. Adjusting the chemical composition and altering the microstructure by processing, as in the present case, are potential routes. A transformation-induced dual-phase high-entropy alloy with high strength and ductility can be produced by metastability engineering. The promotion of strengthening and toughening mechanisms using differing strategies (interstitial alloying elements, dislocation-slip, lowered stacking-fault energy, transformation-induced or twinning-induced plasticity) have been essayed, but the activation of such mechanisms requires considerable optimization of the composition. The high strength of the present processed alloy was attributed to the formation of recrystallized fine grains, while the increase in elongation was attributed to the phase transformation.

A medium-entropy alloy system, CrCoNiSi$_x$ (x = 0.1, 0.2, 0.3) was designed[93] in an attempt to increase both strength and ductility. The alloys were all single-phase, with a face-centred cubic structure. The lattice distortion increased with increasing silicon

Materials Research Forum LLC
https://doi.org/10.21741/9781644903230

content (table 26). All 4 alloys had microstructures with equiaxed and randomly oriented grains containing annealing twins. The volume fraction of annealing twins was estimated from the number fraction of twin boundaries, and was equal to 6.3, 10.2, 13.3 and 16.7% for x = 0, 0.1, 0.2 and 0.3, respectively. There was a monotonic increase in grain size with increasing x. The recrystallization rate and average grain size increased with increasing silicon content. The stacking-fault energy decreased with increasing silicon content. The ultimate tensile strength and uniform elongation of the $CrCoNiSi_{0.3}$ alloy, as compared with those of the Si-free alloy, increased from 790MPa to 960MPa and from 58% to 92% (table 27). The product of UTS and total elongation, used as a sort of goodness-factor, changed from 46 to 88GPa%. The strengthening was attributed to the reduction in stacking-fault energy and to an increase in lattice distortion due to the silicon. More concentrated and thinner deformation twins, plus multiple twinning, were seen in the Si-containing alloys. A nanoscale diffusionless transformation from face-centred cubic to hexagonal close-packed occurred during room-temperature tensile deformation in the silicon-containing alloys, thereby further increasing the work-hardening and uniform elongation.

Table 26. Effect of silicon additions upon the lattice distortion and grain size of $CrCoNiSi_x$

X	Lattice Distortion	Grain Size (μm)
0	2.3×10^3	16
0.1	10.6×10^3	23
0.2	14.4×10^3	51
0.3	17.1×10^3	58

Table 27. Effect of silicon additions upon the tensile properties of CrCoNiSi$_x$

x	Yield Stress (MPa)	UTS (MPa)	Elongation (%)	Product (GPa%)
0	405	790	55	46
0.1	452	856	70	64
0.2	476	926	83	80
0.3	502	960	90	88

A controlled-gradient nanoscale dislocation-cell structure was introduced into a stable single-phase face-centred cubic high-entropy alloy, Al$_{0.1}$CoCrFeNi, in an attempt to increase strength without losing ductility[94]. During straining, the structural gradient introduced the progressive formation of a high density of minute stacking faults and twins which nucleated from the numerous low-angle dislocation cells. Stacking fault-induced plasticity and refined structures, together with accumulated dislocations, contributed to the plasticity and increased the strength and work-hardening. The alloy initially contains randomly-oriented equiaxed fine grains having an average diameter of some 46mm. This alloy was known to have a stacking-fault energy of 6 to 21mJ/m^2. The extremely high density of minute stacking faults, twin nucleation and accumulation-dominated plastic deformation following the initial application of tensile strain was unexpected. By tuning the cyclic torsion (angle of 20° or 6°), it was possible to prepare 2 different gradient dislocation cell structures by using a constant torsion number of 200 cycles. These were designated as being imposed high and low accumulated torsion plastic strains (18.4 and 5.2, respectively, in the upper layer). The high samples reflected the main microstructural features of the hierarchical dislocation structure. Grains from the surface to the core of the high samples were homogeneously distributed, with faceted morphologies, unchanged size and random crystallographic orientations. The same observations were made before cyclic torsion. The notably unchanged grain features following torsion were very different to those of traditional homogeneous, or gradient, nanostructures; with the greatly refined grain-sizes and higher densities of high-angle grain boundaries that are produced by conventional severe plastic deformation. Most of the grains in the sample core exhibited the typical planar single-slip dislocation configuration and a relatively low dislocation density. Well-developed single slip-induced dislocation walls with low-angle boundaries were sometimes observed. Massive low-angle boundaries, with misorientations of up to 15° were introduced into the uppermost grain interior and were distributed such as to become lower in volume fraction and greater in size with increasing distance from the top surface. Similar dislocation structures were observed in the low samples at low cumulative plastic strains, including undeveloped dislocation cells with a greater average size (~450nm) in the uppermost surface. The dislocation cells in the samples were all caused mainly by intensive multi-slip full dislocation interactions within a complex gradient stress/strain state following high cumulative torsion strain. No stacking faults or deformation twins were observed in bulk samples, suggesting that dislocation-controlled plastic deformation occurred during cyclic torsion. Synchrotron X-ray diffraction scanning of the as-prepared high samples revealed a spatial gradient–distributed dislocation density of up to $8.8 \times 10^{14}/m^2$ within the uppermost ~200mm. The gradient dislocation structure was compositionally homogeneous, with no segregation at the cell wall. The hierarchical dislocation cell structure led to greatly improved tensile properties, with tensile tests of the 20° and 6° samples indicating 0.2% offset yield strengths of 362 and 539MPa. A high uniform elongation (42.6%) was found for 20° samples. The latter samples also exhibited steady strain-hardening, with a slightly decreased work-hardening rate; going from 1.28GPa at 3% strain to 0.99GPa before necking. These properties were compared with those of other materials (figure 11).

The gradient structure also led to an unexpected deformation-induced continuous hardening behaviour: due to the sample-level gradient dislocation structure over the entire cross-section, a distinct HV gradient was present – from 3.7GPa at the uppermost surface to 2.2GPa in the central region of 20° samples. In 6° samples, the HV gradient ranged from 2.3 to 1.7GPa. The marked increase in yield strengths was attributed to the nanoscale dislocation cell units with low-angle boundaries. In the case of the uppermost surface layer of the 20° gradient dislocation material, the ultra-high HV value indicated that the massive low-angle dislocation cells were effective in resisting dislocation motion due to their nanoscale size and high dislocation density. At a strain of 3%, there was almost no change in the grain shape, size or orientation, but the density of low-angle boundaries was reduced in the sub-surface layer and in the core. The presence of the dislocation patterns was attributed to poorly-developed or unstable states at quite low cumulative plastic strains. The pattern was essentially unchanged for the uppermost grains, and long parallel lamellar bundles with an average spacing of about 1.7mm were present in most of the uppermost grains. These decreased in number-density with increasing depth. The main components of the lamellar bundles were stacking faults with a few twin boundaries.

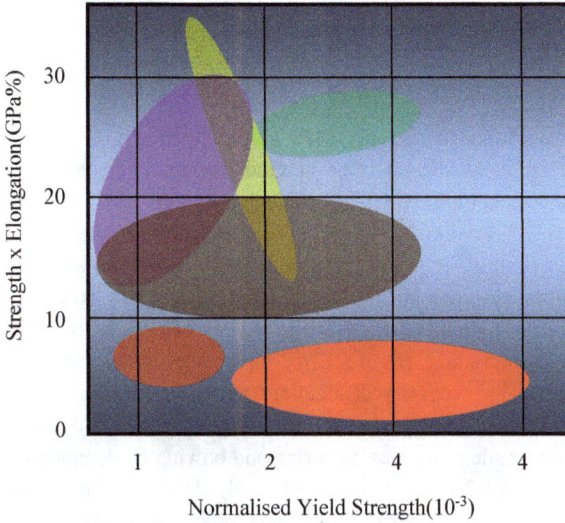

Figure 11. Mechanical properties of the gradient dislocation structured $Al_{0.1}CoCrFeNi$ high-entropy alloy (green), compared with those of other materials, with the product of strength and ductility plotted versus yield strength, normalized by the Young's modulus of the present material. Yellow: gradient-nanograined 316 stainless steel, purple: fine-grained high-entropy alloys, brown: gradient-grained high-entropy alloys, dark red: gradient nanograined Cu/Cu-Al alloy, light red: gradient nanograined nickel

Microstructures having single or dual gradients were created[95] in a $Al_{0.5}Cr_{0.9}FeNi_{2.5}V_{0.2}$ medium-entropy alloy. The non-equiatomic alloy was subjected to surface mechanical attrition, followed by short-term annealing and long-term aging. The dual-gradient samples had an increasing grain-size and a decreasing volume fraction of nano-precipitates ranging from the surface to the centre. The dual gradient had a synergetic strengthening/toughening effect when compared with that of a single gradient and associated precipitation. The nucleation of adiabatic shear-bands was delayed and their propagation was slowed in structures with dual gradients, as compared with structures having single gradients. This resulted in better dynamic shear properties. A greater strain-gradient and higher density of geometrically necessary dislocations were introduced into structures with dual gradients, and this resulted in additional strain-hardening. Higher numbers of dislocations, stacking-faults and Lomer-Cottrell locks could accumulate due to interactions between these defects and B2 or L1$_2$ precipitates, due to the higher volume-fraction of nano-precipitates in the surface layer of structures with dual gradients. This could delay early strain-localization in the surface layer and lead to better dynamic shear properties.

Early strain-localization has to be suppressed in the uppermost nanostructured layers in order to obtain a better tensile ductility in gradient structures. Such structures were produced[96] in a non-equiatomic Al-Cr-Fe-Ni-V high-entropy alloy with a nickel content of about 50% and a Ni/Al ratio of about 5. There were combined gradient distributions of grain-size and precipitate volume-fraction in the depth direction that were produced by means of surface mechanical attrition and aging. The yield strength and uniform elongation were thereby simultaneously improved, as compared to those of structures with only a grain-size gradient (figure 12). More severe strain gradients, and higher densities of geometrically necessary dislocations were present at domain boundaries in structures with combined gradients. This led to greater hetero-deformation induced hardening and better tensile properties. Shearing and bowing hardening mechanisms were associated with L1$_2$ and B2 precipitates, respectively. Higher volume fractions of these phases in the uppermost layers led to higher precipitation-hardening, and compensated for a decreased strain-hardening.

A carbon-nitrogen co-doped interstitial high-entropy alloy, Fe$_{48.5}$Mn$_{30}$Co$_{10}$Cr$_{10}$Co$_{0.5}$N$_{1.0}$, was subjected[97] to cold-rolling to a strain of 1.74, and annealed. Following cold-rolling, the main microstructure comprised mainly nano-grains, nano-twins, hexagonal close-packed laminae and a high density of dislocations. This imparted a hardness of 466.7HV, a tensile strength of 1730MPa and a ductility of 2.44%. The nanostructures and hardness (462.5HV) were retained up to an annealing temperature of 600C. Following annealing (650C, 1h), the ultrafine-grained microstructure contained recrystallized grains with an average size of 0.91μm and nano-precipitates with an average diameter of 90.8nm. The combined effects of ultra-fine grains, nano-precipitates, twinning and solutes led to a strain-hardening of 0.81, a yield strength of 984MPa and a ductility of 20%. The co-doping produced a marked drag on dislocation slip, leading to a nano-scale mean-free-path of dislocation slip of 1.44nm and an activation-volume of 15.8b^3. The addition of interstitial carbon had led to a ductility of 54% and a little-affected yield strength. This was attributed to the effects of twinning-induced plasticity and interstitial solid-solution strengthening. The addition of 1.15at% of nitrogen as an interstitial element markedly increased lattice-friction, leading to an increase of 156MPa in yield strength and of 9.1% in uniform elongation. Dislocation-slip and twinning predominated during tensile deformation, and phase changes completely ceased. Partially recrystallized and deformed

Materials Research Forum LLC

https://doi.org/10.21741/9781644903230

grain structures in the non-doped material could impart a yield strength of 1190MPa and a ductility of 12%.

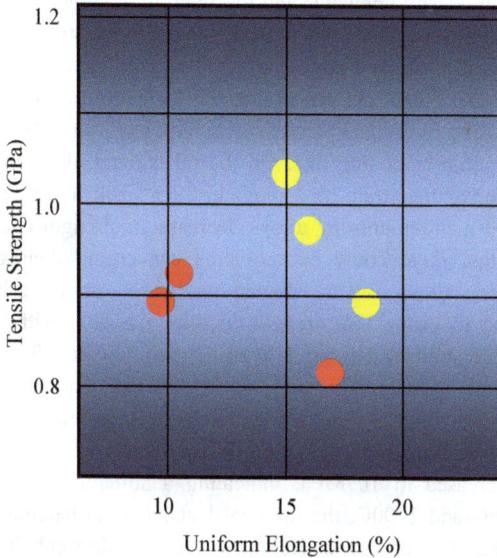

Figure 12. Yield strength versus uniform elongation for double- and single-gradient high-entropy alloys. Red: structures with single gradient, yellow: structures with two gradients

Rolling at room temperature of the face-centred cubic high-entropy alloy, $Al_{0.5}CoCrCuFeNi$, produced a nanocrystalline structure and a good combination of strength and ductility[98]. The yield strength and ultimate tensile strength were 1284 and 1344MPa, respectively, with an elongation of 7.6%. Following annealing (900C, 600s) the elongation doubled to 15.3%, with an increase of some 20% in strength. This combination of strength and ductility was attributed to quasi-dynamic recrystallisation during cold-working and limited grain-growth during annealing.

The $Ni_2CoFeV_{0.5}Mo_{0.2}$ medium-entropy alloy, with a single-phase face-centred cubic crystal structure, exhibits a high work-hardening ability, yield strength, ultimate tensile strength and uniform elongation at up to 800C[99]. An increase in the strain-hardening rate occurs following the elastic-plastic transition during tensile deformation at 25 to 800C. A high density of dislocation forests and solute atmospheres in the concentrated solid solution is assumed to account for the increase. The temperature effects upon the mechanical properties are closely related to dislocation structures, dislocation densities and dynamic strain-aging. Upon increasing the deformation temperature from 25 to 800C, the dislocation density undulates and the fraction of screw dislocations increases. A strong solute-pinning, together with a predominant forest-strengthening mechanism, negative strain-rate sensitivity and large activation volume increase the strain-hardening rate at 400 to 700C. At 800C, the predominant deformation mechanism changes from forest-dislocation cutting to dislocation cross-slip, with increasing strain. This results in a sudden increase in strain-hardening rate. On the other hand, dynamic strain-aging and plastic instability impair the strain-hardening. In later work it was again noted[100] that almost all high- and medium-entropy alloys decrease in strength and ductility at high temperatures, and that there could be problems concerning their oxidation at high temperatures. Some alloys were singled-out. The multi principal-component compositions helped to evade the strength-ductility paradox, with $Ni_2CoFeV_{0.5}Mo_{0.2}$ exhibiting appreciable strain-hardening as well as good ductility from 25 to 800C but, when the temperature was greater than 800C, it exhibited strain-softening and a considerably reduced ductility. As the temperature was increased from 25C to 900C, the yield strength decreased from 310 to 191MPa. When the temperature increased to 1000C, the yield stress decreased to 107MPa, indicating a softening effect with increasing temperature. At 900C and 1000C, the material lost its strain-hardening ability and the ultimate tensile strength decreased. Here the alloy underwent immediate necking following yielding, leading to negligible uniform plastic deformation. As the temperature increased from 25 to 800C, the uniform elongation decreased from 62% to 32%. As the temperature increased from 25 to 1000C, the failure-strain decreased from 73% to 47%.

High-entropy alloys with the composition, $CoFeMnNiMo_{0.2}Al_x$, where x was 0, 0.3, 0.4 or 0.5, were prepared[101] from 3N5-purity elements. The $CoFeMnNiMo_{0.2}Al_{0.5}$ alloy, with a face-centred cubic plus $L1_2$ plus body-centred cubic microstructure exhibited a compressive strength of 1869MPa and a fracture strain of 39.6%, with a nano-hardness of almost 8GPa. The compressive tests were performed at a strain-rate of 0.0005/s, and the hardness tests were performed using a load of 500g for 20s. The $CoFeMnNiMo_{0.2}$, with face-centred cubic structure, did not fracture and had a high ductility. Its compressive yield strength was 288MPa. As the aluminium content was increased, the yield strength increased from 829MPa for x = 0.3 to 1869MPa for x = 0.5. The aluminium additions did not cause severe brittleness. The fracture-strains of the x = 0.3, x = 0.4 and x = 0.5 were 57.6%, 42.1% and 39.8%, respectively. As x increased from 0 to 0.5, the volume fraction of the body-centred cubic phase increased from 0 to 35%. The size of the precipitates in the x = 0.3 alloy was 8.9nm, and so the improvement in yield strength was attributed to a shearing mechanism. For the x = 0.3 and x = 0.5 alloys, strengthening was attributed mainly to precipitation-hardening. Due to their relatively small size and distribution, shearing was more important than Orowan strengthening.

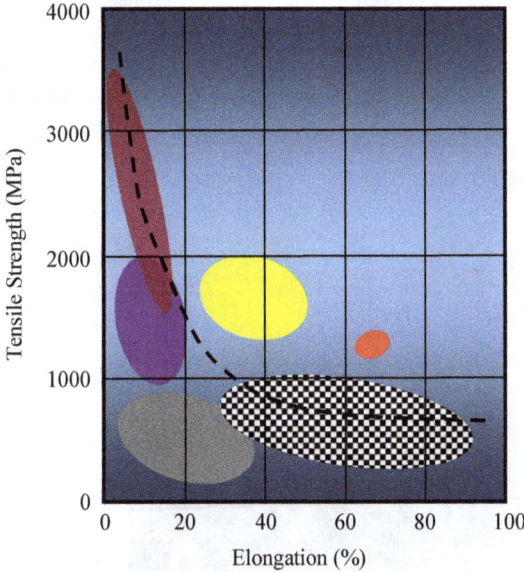

Figure 13. Ultimate tensile strength and ductility for $Cr_{26}Mn_{20}Fe_{20}Co_{20}Ni_{14}$ high-entropy alloy at 4.2K. Light red: present results, chequer: copper alloys, light gray: aluminium alloys, purple: titanium alloys, yellow: stainless steels, dark red: NiFe

A face-centred cubic non-equiatomic $Cr_{26}Mn_{20}Fe_{20}Co_{20}Ni_{14}$ high-entropy alloy with a stacking-fault energy of 17.6mJ/m^2 was prepared[102] by vacuum induction-melting, forging and annealing. The annealed material had a fully recrystallized microstructure, with grain-sizes ranging from 10 to 100μm, and a single-phase microstructure. The recrystallized material exhibited a good combination of strength and ductility between 4.2K and 293K. When decreasing the test temperature from 293K to 77K, the ductility and ultimate tensile strength gradually increased by 30%, to 95%, and by 137%, to 1020MPa. At 4.2K, the ductility remained at 65% while the UTS increased by 200%, to 1300MPa. Microstructural analyses revealed a change in deformation mechanism, from dislocation-slip and nano-twinning at 293K, to nano-phase transformation at 4.2K. The temperature-dependence of the strain-hardening rate exhibited 3 distinct stages, with a broad steady-state which was different to the monotonic decrease found in conventional polycrystalline metals. The rapid drop in strain-hardening rate in stage-I at a given temperature corresponded to the conventional transition from elastic to dislocation-slip dominated plastic deformation. Cooperation and competition between multiple nano-twinning and nano-phase transformation were concluded to be responsible for the good tensile properties which were observed at cryogenic temperatures and compared well with those of other materials (figures 13 and 14).

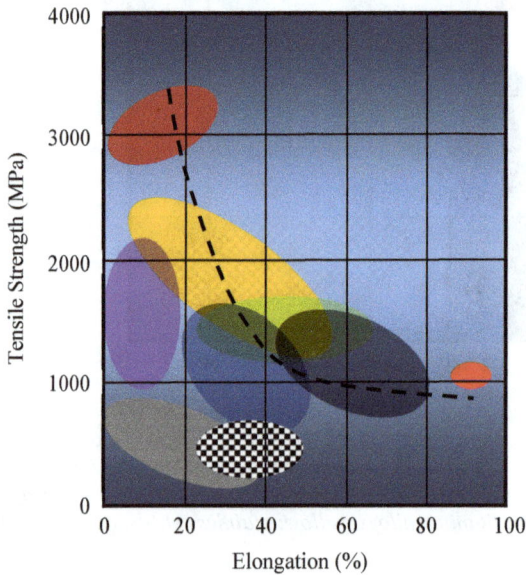

Figure 14. Ultimate tensile strength and ductility for $Cr_{26}Mn_{20}Fe_{20}Co_{20}Ni_{14}$ high-entropy alloy at 77K. Light red: present results, chequer: copper, light gray: aluminium alloys, purple: titanium alloys, dark gray: high-entropy alloys, blue: medium-entropy alloys, yellow: stainless steels, dark red: NiFe

Composites

Aluminium-based

Dual-matrix carbon-nanotube composites were prepared[103] in which composite particles of aluminium, reinforced with carbon nanotubes, were embedded within a soft aluminium matrix. This approach combined the high strength of the nanotube-reinforced region with the ductility of the soft aluminium matrix. The quality of the interface was essential in promoting bonding between the dissimilar components. Poor interfacial bonding led to deterioration of the composite and rendered it impossible to evade the strength-ductility paradox.

Stir-cast metal-matrix composites were prepared[104] by reinforcing AA2024 with boron carbide particles, having mesh-sizes of 100, 200 or 300, in quantities of 1, 3 or 5wt%. There was a uniform distribution of the boron carbide in the AA2024. Tensile specimens were solutionized (520C, 24h, water-quench) and aged (175C, 1, 2, 3 or 5h). The ductility decreased with increasing percentage weight of boron carbide (table 28), while the yield strength (table 29) increased. The latter increase was attributed to the increase in the fraction of B_4C particles and bonding between the reinforcement and the matrix. The ductility of the composites decreased with increasing amount of reinforcement. Use of the Taguchi technique showed that the main factors which affected the ductility of the composites were the percentage weight of B_4C, the mesh size and the aging-time … in decreasing order. The greatest percentage contribution (50.27%) was the percentage weight of B_4C while the aging-time made the lowest contribution (15.28%). The main factors which affected the yield strength of the composites were the aging-time, mesh size and percentage weight of B_4C … in decreasing order. The greatest contribution (60.36%) was that of the aging-time while the percentage weight of B_4C made the lowest contribution (6.64%). The reinforcement particles led to a dimpled ductile fracture.

Table 28. Ductility of AA2024-B$_4$C composites

B$_4$C(wt%)	Aging(h)	Mesh Size	Ductility(%)
1	1	100	9.17
1	3	200	10.84
1	5	300	10.24
3	1	200	9.54
3	3	300	9.81
3	5	100	7.29
5	1	300	9.24
5	3	100	6.81
5	5	200	6.54

Table 29. Yield strength of AA2024-B$_4$C composites

B$_4$C(wt%)	Aging(h)	Mesh Size	Yield Strength(N/mm^2)
1	1	100	178.82
1	3	200	184.25
1	5	300	194.41
3	1	200	178.78
3	3	300	186.35
3	5	100	220.64
5	1	300	158.35
5	3	100	220.35
5	5	200	215.82

Materials Research Forum LLC
https://doi.org/10.21741/9781644903230

Composites of AA6016, reinforced with 1 or 2vol% of Al_3Zr, Al_2O_3 and ZrB_2 nanoparticles, were prepared[105] by *in situ* chemical reaction in an electromagnetic field and solidified under pressure. As well as the above particles, Fe,B-rich and AlB_2 particles also formed. Under squeezing pressures of 10 to 30MPa during solidification, fine Mg_2Si precipitates formed in the matrix of 1vol%-reinforced composites while none formed in the matrix of 2vol%-reinforced composites. The grain size of the squeezed 1vol% composite decreased from 102μm to 58μm, while that of the squeezed 2vol% composite decreased from 83μm to 50μm. The composites, squeezed under high pressures, exhibited a both a high strength and a high ductility, together with a lower strain-hardening effect in the aged state. Unlike AA6016, which has a high strength but loses much of its ductility during aging due to increased strain-hardening, the simultaneously improved tensile strength and ductility of the squeezed composites were attributed to grain-boundary strengthening, to dislocation-strengthening, to Orowan strengthening and to an amorphous layer at the α-Al/θ-Al_2O_3 interfaces.

It was shown[106] that rapid solidification during laser additive manufacturing could arrange the engulfing of reinforcing particles by aluminium grains, thus helping to separate the stress-concentrations caused by grain boundaries and added reinforcements. Experiment indicated that the intragranular dispersion of particles inhibited crack-initiation and induced strain-hardening, leading to markedly improved mechanical properties. The present method thus led to a notably improvement in tensile strength and ductility. The properties of a selective laser melted Al-TiB_2 composite were compared with those of a conventional powder metallurgy sample having the same particle content of 15vol%. The selective laser melted composite exhibited a 2.8-fold increase in total elongation, from 3.1% to 8.6%, and a 30% improvement in the tensile strength, from 176MPa to 230MPa, over the equivalent values for the powder metallurgy sample. Grain-refinement played a minor role in imparting the higher strength, given that the grain sizes were similar: 20μm for the laser-melted and 15.7μm for the powder-metallurgy material. The Young's modulus of the former composite was 16% higher: 109MPa as opposed to 91GPa. This suggested that the uniform intragranular dispersion of TiB_2 particles improved the load-bearing ability. The fact that the matrix was high-purity aluminium emphasized the role played by the intragranular dispersion. Dislocation-blocking and accumulation were observed near to the embedded TiB_2 particles, leading to appreciable strain-hardening. Fracture-surface analyses showed that the failure mode changed from interfacial de-bonding, in the powder metallurgy composite, to ductile failure in the selective laser melted composite; with no crack-initiation at interfaces. In the latter composite, intragranularly-dispersed TiB_2 reduced the stress concentration at grain boundaries and permitted extensive plastic deformation of the aluminium matrix ahead of crack nucleation. The presence of local misorientations near to intragranular particles, and the size of local so-called domains increased with applied strain. Particles in the matrix could be treated as being Eshelby inclusions, where the elastic stress-field increases with particle size, so that intragranular TiB_2 particles not only accumulated dislocations but also created relatively large elastic stress/strain fields. This then led to the accumulation of geometrically necessary dislocations and to an associated extra back-stress hardening. The extensive geometrically necessary dislocation-distribution interacted with gliding dislocations and promoted marked strain-hardening. The present intragranular dispersion strategy thus satisfied two important criteria for the ductility improvement of metal composites: a high strain-hardening capacity and a high fracture resistance. In conventional metal composites the particles would be distributed at grain boundaries, and the localisation of geometrically necessary dislocations at grain boundaries would intensify local stress concentrations and result in early cracking-onset and failure. Localized geometrically necessary dislocations at grain boundaries also reduced the interplay between dislocations, and consequently limited dislocation multiplication and accumulation. Composites with intergranularly distributed particles then offered limited strengthening, low strain-hardening and low ductility, but the present composite offered a 30% increase in tensile strength plus a tripled ductility.

The Mg-Ag micro-alloying of Al-4.5%Cu/5%TiB_2 composites via *in situ* molten-salt reaction and heat treatment was studied[107]. Uniformly dispersed TiB_2 particles and semi-coherent grain boundaries between the α-Al matrix and reinforcement phases such as TiB_2, Al_2Cu and $Al_5Cu_6Mg_2$ facilitated dislocation motion and eliminated stress concentrations, thus potentially improving the strength-ductility compromise. Tensile testing at room temperature identified 3 types of composite. Modified processed composites exhibited a superior combination of tensile strength and high uniform ductility, with the yield strength attaining a value of 401MPa, and the ultimate tensile strength attaining 497MPa. These were improvements of 15.6% and 10.2% when compared with data for conventionally processed composites. The modified processed composites also exhibited a high tensile ductility, with a uniform elongation of 6.1%. This was more than twice than that (2.8%) of conventionally processed composites. In the case of conventionally processed Al-4.5%Cu/5%TiB_2 composites, the microstructures consisted mainly of refined α-Al grains with an average size of 161μm, with Al_2Cu phase at the grain boundaries. Most of the TiB_2 particles, with irregular profiles, adhered to one another to form agglomerates. The boride could potentially lead to large-scale grain refinement, but the sub-micron size of the particles and large surface-area/volume ratios instead led to van der Waals and adhesion forces causing the particles to aggregate into clusters. These agglomerated boride particles could then cause stress concentrations and become crack sources during deformation. When 0.6% of magnesium and 0.1% of silver were added, boride agglomerations persisted in the matrix but the average size of α-Al grains decreased to 82μm. The segregation of large boride agglomerates was markedly improved because the surface-active element (magnesium) reduced the work-of-adhesion, weakened the Al/TiB$_2$ interfaces and inhibited boride agglomeration. Magnesium could also reduce the interfacial energy of the entire system because of its higher oxygen-affinity: magnesium reacts with Al$_2$O$_3$ to form MgO or MgAl$_2$O$_4$, reducing the amount of Al$_2$O$_3$ on the melt surface and improving the wettability of melt to boride particles and inhibiting the segregation of boride clusters. The combination of Mg-Ag alloying and heat treatment had a marked effect upon the sizes of α-Al grains, with the microstructure consisting of fine columnar and equiaxed grains, and the mean size of α-Al grains being 66μm. The yield strength of the composites increased as the α-Al grain size decreased. Finer grain structures reduced flaw-sizes and made more grains available to support deformation, increasing the resistance to crack propagation and improving ductility. At a strain of about 1%, dislocation slip was activated around boride particles; the early stages of planar dislocation slip across the grain boundaries. Upon increasing the strain to 3%, more dislocations were activated and in much higher numbers. Upon increasing the strain to 6.2%, hexagonal dislocation networks formed. This was expected to lower the defect energy and increase ductility. The good mechanical properties of the modified composites were attributed to various contributions. Boride nanoparticles in low concentrations could be easily engulfed during solidification, thus greatly limiting the agglomeration of boride particles. Numerous σ-Al$_5$Cu$_6$Mg$_2$ and Ω-Al$_2$Cu phases were precipitated during multi-step heat treatment after adding the trace amounts of magnesium and silver. The dislocations rearranged and transformed into dislocation networks and, together with the recrystallized and uniform microstructure, encouraged homogeneous deformation.

Copper-based

The strength-ductility paradox can also be a feature of copper which is reinforced with carbon nanotubes, where the interface characteristics of the nanotubes are critical factors determining the mechanical properties of copper-matrix composites. A new strategy of surface and intratube decoration of the carbon nanotubes was instituted[108] in order to evade the paradox. By decorating the surface of the nanotubes (inner diameter: 20nm, outer diameter: 30nm, length: 0.5 to 2μm) with CuO nanoparticles, and exploiting the good wettability between CuO and the copper matrix, a uniform dispersion of carbon nanotubes was achieved by ball-milling. The oxide provides good wettability with the copper matrix and can normally be deposited onto the surface of carbon nanotubes via molecular-level mixing. This method is not suitable for preparing composites in large quantities, hence the use of high-energy ball milling. By combining the two methods, the nanotubes could first be decorated with CuO nanoparticles and then ball-milled with copper powder. In the final stage, CuO nanoparticles on the surface of carbon nanotubes were reduced to copper, not only maintaining the integrity of the nanotubes, but strengthening the interfacial bonding by creating Cu–O bonds between the copper matrix and the carbon. High-density interface dislocations and interfacial disordered areas were formed between the copper matrix and the carbon nanotubes so as to form a strong interfacial bond. By decorating the inner walls of carbon nanotubes with copper nanoparticles, the interfacial shear stress between the copper matrix and the carbon nanotubes was improved due to the extrusion effect of copper nanoparticles on the inner walls. The copper within the tubes could moreover reduce the intra-tube resistivity of the nanotubes by increasing their conductive cross-section. As a result, a composite with a strength of 272MPa and a ductility of 14.3% was created. As a bonus, the conductivity was 93.6%IACS.

Materials were prepared[109] which comprised carbon nanotubes that were reinforced with copper composite foams. Nickel nanoparticles were used to decorate the nanotubes and thus improve the interfacial bonding. The transition area between the nanotubes and the copper tended to be populated with Ni_3C. The as-received nanotubes had clear smooth surfaces, regular walls and hollow channels. The nickel-treated nanotubes exhibited a uniform distribution of nickel nanoparticles with a mean size of 3 to 5nm on their surfaces. Nickel deposition reduced the surface energy of the nanotubes and prevented their agglomeration while improving the wettability and interfacial compatibility. The copper matrix could also form a complete solid solution with the decorated nickel nanoparticles. The hardness of the pure copper was 101.9 to 108.3HV while that of the composite-skeleton region was 128.1 to 131.4HV. The hardness increased linearly with the addition of nanotubes. The tensile strength (table 30) of the composite could be regarded as being made up of the basic pure copper strength, plus additions due to grain refinement and load-transfer (table 31). The latter was a major contributor to the overall strength. By modifying the surface of carbon nanotubes, the load transfer could be increased from 9.8MPa to 32.0MPa. Partial reaction of the nanotubes with the interface led to better load-transfer.

Table 30. Tensile properties of copper/carbon-nanotube composites

Material	Nanotubes	UTS(MPa)	e_f(%)
Ni-CNT/Cu$_f$Cu	0.04wt%	364.9	40.6
CNT/Cu$_f$Cu	0.04wt%	334	40.5
CNT/Cu-Ti	0.4wt%	355	22.8
TiC-CNT/Cu	1.5vol%	281.0	20.1
Ni-CNT/Cu	0.5vol%	292	34
Cu/CNT	0.5vol%	275	24
Cu-CNT/Cu	0.4wt%	272	14.3

Table 31. Contributions to the strength of copper/carbon-nanotube composites

Material	Matrix(MPa)	Grain Boundaries(MPa)	Load-Transfer(MPa)
Ni-CNT/Cu$_f$Cu	310.6	22.3	32.0
CNT/Cu$_f$Cu	310.6	13.6	9.8

In similar work, copper-matrix composites reinforced with carbon nanotubes were studied[110] which had a 3-dimensional skeleton. Open-cell composite foams with uniformly embedded nanotubes were filled with a dense copper phase by spark plasma sintering so as to form inhomogeneous structures. The materials exhibited a tensile strength of 366.7MPa, combined with a fracture elongation of 31.78% (figure 15), at a modest nanotube content of 0.2414vol%. When the nanotube content was less than 0.3221vol%, the tensile strength of the composite increased with increasing nanotube content. When the nanotube content was 0.3221vol%, the tensile strength was 281MPa and the elongation (24.25%) of the composites markedly decreased. Excess amounts of nanotube led to such agglomeration as to form pores in the structure, thus resulting in poor interfacial bonding and creating crack-sources. The strengthening effect of the nanotubes was again attributed to their high load-transfer efficiency. The pure copper region underwent appreciable plastic deformation during tensile straining and helped to regulate the plasticity of the composite. A diffused interface layer strongly impeded sliding between the nanotubes and the copper matrix.

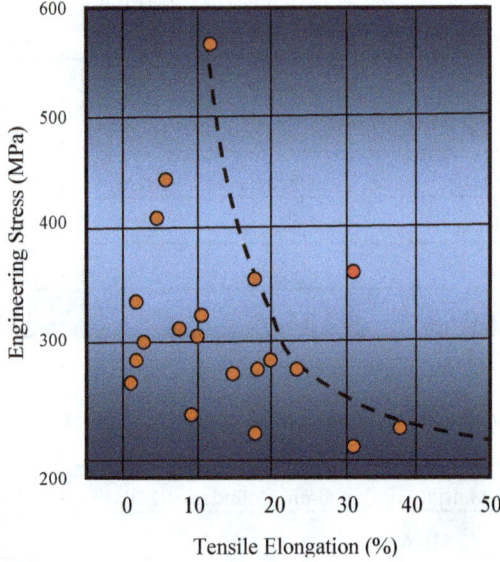

Figure 15. Tensile strength and elongation of copper/carbon-nanotube composites. Orange points: previous carbon nanotube reinforced metal-matrix composites. Red point: present composite.

Open-cell copper-matrix foams containing uniformly distributed reduced graphene oxide were prepared[111] using electrodeposition so as to create a 3-dimensional skeleton-reinforced copper composite. The foam pores were filled with a pure copper phase via spark plasma sintering. Nanoparticles of Cu_2O were formed at the graphene/copper interface and improved the interfacial bonding. The composites combined a tensile strength of 343MPa with an elongation-to-fracture of 39.4%, for a reduced graphene oxide content of 0.024wt%. Composite interfaces are generally distributed throughout the composite volume and greatly affect dislocation movement, thus resulting in a decreased plasticity; hence the choice of a skeleton-reinforced structure in the present case. The main strengthening mechanisms were grain-refinement, Orowan looping, dislocation-strengthening due to thermal mismatch and load-transfer. The electrical conductivity was also high (93.24%IACS).

Iron-based

Heat treatment could be used to improve the properties of $Fe_{0.6}MnNi_{1.4}$-5vol%TiC composites[112]. Following annealing (1100C), precipitates of $M_{23}C_6$, where M was iron, manganese or nickel, were formed, leading to a yield strength, an ultimate tensile strength and ductility of 307.4, 805.5MPa and 51.1%, respectively. An orientational relationship existed between the matrix alloy and the nano-sized carbide precipitates. The introduction of the ceramic particles had generally increased the strength of medium-entropy alloys: the ultimate tensile strength of $FeCoNiCu_{2.0}$-10volTiC% composite was 47.6% higher than that of the matrix alloy, and the yield strength of CoCrNi alloy with 2wt% of SiC had been increased from 352 to 595MPa while the ductility was reduced from 53.6 to 18.6%. The addition of TiC particles had increased the strength of $Fe_{0.6}MnNi_{1.4}$ alloy by 15.8% and reduced the elongation.

Titanium-based

Multilayer nanocrystalline thin films were studied[113] which consisted of relatively thick porous TiN ceramic layers alternating with quite thin (0.8nm to 34nm) titanium layers, with a constant thickness ratio of about 17.5. As the metal thickness decreased, the overall deformation mechanism of the multilayer films split into a dislocation-dominant confined layer slip mechanism in the metal and a diffusional creep-dominant process in the porous ceramic layer when the metal thickness was 6.7nm. This combination of mechanisms led to the highest strain-hardening rate due to the confined layer slip mechanism in one layer, at the same time as the highest strain-rate sensitivity value due to diffusional flow in the other layer, thus evading the strength-ductility paradox.

A titanium-matrix composite was reinforced[114] with TiB nano-whiskers with a reticulated microstructure. It was prepared from a composite powder by means of selective laser melting. The composite powder had a micron-scale reticulated structure, with micro-nano TiB whiskers lying along the grain boundaries. This was further reduced to the nano-scale during the laser-treatment, due to the rapid solidification rate. The composite had a strength of 851MPa and a ductility of 10.2%. The smaller particles, with their specific interfacial area, had had the effect of increasing the strength while retaining a good ductility.

Uniformly-nitrided, layer-structured and gradient layered-structure titanium composites were prepared via *in situ* laser powder-bed fusion[115]. Depending upon the nitrogen concentration, the microstructure of the commercial-purity titanium ranged from coarse α-Ti grains to fine acicular martensitic α′-Ti. No TiN was formed when the nitrogen concentration was below 10%. The fractions of TiN for nitrogen concentrations of 10% and 15% were 2.2% and 2.7%, respectively. The hardness and strength of uniformly-nitrided composites, with an overall distribution of TiN, increased with increasing nitrogen concentration and the plasticity decreased. When the TiN was distributed as layers, the composites exhibited balanced strength and plasticity for various ratios of nitrogen concentration. The ultimate tensile strength ranged from 830 to 1100MPa while the plasticity ranged from 9 to 30%. The use of nitrogen for the *in situ* synthesis of TiN layers via laser powder-bed fusion in a cross-gradient layer-pattern arrangement permitted the appearance of higher back-stresses in the soft titanium layer, than those in a single-layered structure, before the hard TiN layer began to yield. This then led to a better strength-ductility compromise in gradient layer-structured titanium composites. In the case of laser powder-bed fused pure titanium or alloys, solid solutions and brittle precipitates could be caused by a high content of interstitial nitrogen and oxygen and resulted in degradation of the plasticity. Samples of commercial-purity titanium, with low nitrogen concentrations, were therefore satisfactory, with UNTi5 exhibiting both high strength and ductility (table 32). The ultimate tensile strength and yield strength of uniformly-nitrided UNTi5 were 37.7% and 47.9% higher than those of nitrogen-free commercial-purity titanium, respectively. A comparison of UNTi10 and UNTi15 showed that the strengths were only slightly improved by the high nitrogen concentration, while the plasticity was essentially zero. Interstitial nitrogen could greatly increase the transformation temperature of high-temperature prior β grains, expand the α-grain region and promote martensite formation, such as to improve the strength of uniformly-nitrided composites. When the nitrogen concentration was above 10%, the UNTi10 and UNTi15 samples contained more nitrogen solid solution, plus some excess TiN particles, giving the titanium matrix a low ductility and decreasing the mechanical properties.

Materials Research Forum LLC
https://doi.org/10.21741/9781644903230

Table 32. Tensile properties of laser powder-bed fused commercial-purity and uniformly-nitrided titanium composites

Sample	UTS(MPa)	Yield Strength(MPa)	Elongation(%)	E(GPa)
CPTi-Ar	696.48	599.19	33.60	107.71
UNTi5	958.83	886.17	17.27	115.01
UNTi10	1006.60	907.64	2.37	119.10
UNTi15	1132.39	1032.29	1.03	120.16

When there was a soft titanium layer present, having the same thickness, the strength increased and the plasticity decreased with the nitrogen concentration. One-dimensional layer-structured titanium and gradient layer-structured materials having nitrogen contents which varied in the Z-direction were prepared using nitrogen concentrations of 5%, 10%, 15% or 20%. Two-dimensional gradient layer-structured materials were formed by varying the nitrogen concentration along the Y- or X-directions, as well as effecting a thickness-change in the build direction. A commercial-purity titanium layer was then added by adjusting the nitrogen concentration along the Y- or X-direction on the first nitrogen interlayer before switching back to an argon rather than a nitrogen atmosphere (indicated by figures such as 0.42[mm] in the table). Varying the nitrogen concentration along the X or Y direction as the commercial-purity titanium matrix built up produced combined-layer materials with gradient changes occurring in the X-, Y- and Z-directions (table 33).

Table 33. Tensile properties of laser powder-bed fusion in situ synthesized layer-structured titanium composites

Sample	UTS(MPa)	Yield Strength(MPa)	Elongation(%)	E(GPa)
LSTi5-0.21	859.46	782.15	29.92	110.47
LSTi10-0.21	904.82	816.92	26.13	117.53
LSTi15-0.21	947.32	861.23	19.58	112.23
LSTi20-0.21	1105.10	1020.69	9.10	115.28
LSTi10-0.30	868.78	793.34	31.98	109.66
LSTi15-0.30	896.46	820.41	26.17	115.05
LSTi20-0.30	1094.59	923.55	16.00	114.96
LSTi10-0.42	827.83	722.57	32.96	106.38
LSTi15-0.42	861.18	774.57	27.72	114.66
LSTi20-0.42	943.16	859.83	20.76	117.15
GLSTi-Z10	899.37	809.44	28.94	108.95
GLSTi-Z15	916.15	837.33	25.02	116.59
GLSTi-YZ	911.28	840.20	24.85	112.71
GLSTi-XZ	923.71	833.74	20.64	114.61
GLSTi-XYZ	917.72	830.05	28.14	115.61

A graphene-reinforced Ti-6Al-4V composite possessing both high strength and good ductility was produced[116] by means of powder metallurgy. Ti-6Al-4V powder with an average diameter of about 100μm was milled with graphite balls at 30Hz for 5h. This exfoliated graphene from graphite balls, and the graphene adhered to the metal powder. The graphene content in the mixed powders was 0.1wt%. The resultant graphene network, with its elastic modulus of some 1TPa and UTS of about 130GPa, closely interacted with the matrix due to the *in situ* generation of interfacial TiC. Grain-refinement, dislocation-blocking and the restraint and bridging of cracks were promoted by the embedded graphene network. This improved the strength and ductility of the composites, which exhibited a better ductility than that of Ti-6Al-4V but also displayed more typically brittle fracture surfaces. The introduction of the graphene network changed the fracture mode of the composite from the transgranular fracture of Ti-6Al-4V to a mixture of intergranular and quasi-cleavage fracture. The former was attributed mainly to brittle fracture of the TiC shell. The propagation of cracks along the cleavage plane of the matrix, together with the blocking of cracks by the graphene network, led to the quasi-cleavage fracture. The restraint of crack propagation due to dislocation-network interactions and crack-bridging of the network, delayed the failure of the composite and thus enhanced the ductility. The yield strength and ultimate strength of a composite with 0.1wt% graphene were improved by 18.72% and 9.31%, while the ductility was improved by 8.66%, thus evading the strength-ductility paradox.

A titanium-alloy (Ti-6Al-4V) matrix composite was reinforced[117] with high aspect-ratio whiskers which had been prepared by the shell-layer encapsulation of a boron-containing polymer precursor, and consolidation by spark plasma sintering. Three boron sources of various sizes were used: 2μm-sized TiB_2, 100nm-sized B_4C and B-modified polysilazane. These reacted with 42μm spherical Ti-6Al-4V particles to produce a network of distributed TiB whiskers. The latter, with aspect-ratios of up to some 304, imparted a strength of 1318MPa together with a ductility of 5.34%. This was attributed to fine grains and to a high load-transfer efficiency to the titanium matrix. In addition, the high aspect-ratio TiB whiskers could bend, unlike particles, without suddenly fracturing or exhibiting limited ductility.

Zirconium-based

A multi-layered zirconium-based ($Zr_{39.6}Ti_{33.9}Nb_{7.6}Cu_{6.4}Be_{12.5}$) bulk metallic glass composite with a controlled volume-fraction gradient of crystalline dendrites was prepared[118] by 6kW-laser additive manufacturing (spot-diameter of about 2.5mm). The latter permitted local control of the cooling rate and the resultant microstructure. Powders ranging in size from 40 to 120μm were used. The cooling-rate of 1000 to 10000K/s permitted vitrification and the creation of gradient composites comprising multilayers with a volume-fraction of crystalline dendrites ranging from 20% to 65% by controlling the cooling rate. The composite had a yield strength greater than 1.3GPa and a tensile ductility of about 13%. The good strength-ductility combination was attributed to a strengthening that resulted from the interaction of adjacent layers and an asynchronous deformation-mode which was associated with the heterogeneous microstructure. Due to the absence of defects such as dislocations and grain boundaries, the bulk metallic glass already possessed a high strength and elasticity. Upon yielding however, plastic strain localizes into a marked shear-band which rapidly propagates and causes catastrophic failure. In order to avoid this, the second phase was introduced. The dispersed soft dendrites promoted the nucleation of multiple shear-bands, impeded the rapid development of shear-bands into cracks and thus improved ductility. The mechanical properties were attributed to sequential plastic deformation and fracture processes, and to a competition between dislocation-motion induced strain-hardening of the crystalline dendrites and shear-banding induced strain-softening of the glassy matrix.

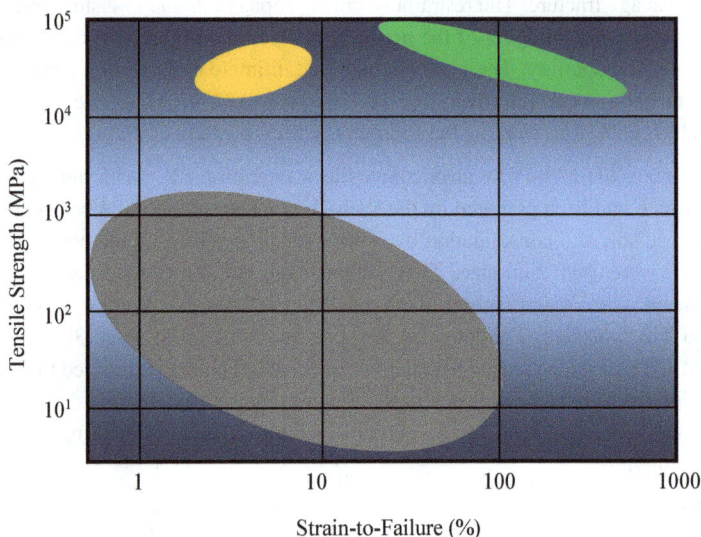

Figure 16. Tensile strength versus strain-to-failure of graphene-based materials. Gray: experimental data on graphene-based materials having multilevel structures and numerous defects, yellow: graphene-based materials with optimised interfacial cross-linking, green: self-folded graphene-based materials

Miscellaneous Materials

The strength-ductility paradox also exists in materials which are far different to the pure metals for which it was originally defined: it appears in carbon-based materials such as graphene[119]. Carbon-based biological materials, such as silk, ironically also suggest paths to making practical materials which fully exploit the strength of graphene. It was suggested that ultra-high ductility graphene-based materials with still-acceptable tensile strength could be created from self-folded graphene sheets. The improved mechanical properties of such materials (figure 16) was here due to the shearing, sliding and unfolding of the self-folded interface. The self-folded length and interface interactions thus governed the strength, ductility and failure of the structures, with that interfacial shearing, sliding and unfolding dissipating mechanical energy. An overall topological strategy was used to improve the properties of carbon nanoscale-structured materials, as guided by molecular dynamics simulations and continuum-model analyses. The optimum interface interactions and self-fold lengths which would maximize the mechanical parameters were predicted. In particular, the strength and ductility of self-folded graphene-based materials could be tailored by choosing the number, n, of self-folded layers with the original thickness of material with n layers being some n times that of pristine graphene. The n self-folded layers eventually unfolded into a single layer, and then that layer unfolded at the end of stretching. The tensile strength of n-layer self-folded graphene-based materials could then be predicted to be σ/n, where σ was the tensile strength (some 100GPa) of monolayer graphene. The fracture strain of n-layer self-folded graphene-based materials meanwhile was approximately $n(1+\varepsilon)-1$, where ε was the fracture strain (some 19%) of monolayer graphene. Simulations showed that increasing n decreased the tensile strength but increased the fracture strain (figure 17); thus implying an optimum n-value of about 3 for combining strength and ductility. The stability of the self-folded materials was checked by simulating (300K, 1ns) samples having various self-folded lengths, L, and interface interactions, ϵ. The stability increased with increasing L and ϵ, indicating that the stability could be controlled by modulating these geometrical and mechanical parameters. The deformation of a sample was typically a 4-stage process. In the first stage, the stress increased rapidly with increasing strain until a first peak stress appeared which corresponded to unfolding of the structure. The increasing tensile load was here transferred into shear-deformation of the self-folded interface. In the second stage there was a slight fall in stress which was attributed to a transition from static to dynamic friction. The stress then changed slightly to a plateau regime. The stress then decreased with increasing strain while interface-sliding markedly reduced the self-folded length, L, constituting a negative-stiffness regime. The elongation in this range of negative stiffness was anticipated to jump if it were loaded at constant force rather than constant displacement. In the third stage, as the strain continued to increase, the stress decreased suddenly due to a critical transition from folded multilayers to an unfolded monolayer occurred. In the fourth stage, the mechanical behaviour was that of completely unfolded layers and thus similar to that of monolayer graphene with its tensile strength of some 100GPa and thickness of circa 0.34nm. A second peak stress was observed upon failure of the unfolded monolayer. Because the original thickness of material with 3-folded layers was some 3 times that of pristine graphene, the tensile strength was expected to be about 33GPa; as observed in simulations. The fracture strain was expected to be up to 258%. Simulations of material with L = 38.4, 51.2, 64.0 or

76.8nm showed that the mechanical properties first increased, and then converged to limited values with increasing L. The critical value of L at which the first stress-peak converges was defined to be the optimum self-folded length. That optimum length decreased with increasing ϵ.

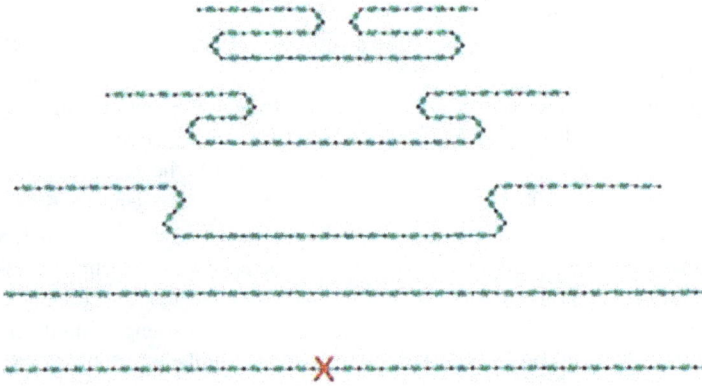

Figure 17. Stages in the nanoscale deformation of self-folded graphene-based material. Taken from Bio-inspired self-folding strategy to break the trade-off between strength and ductility in carbon-nanoarchitected materials, Jia X., Liu Z., Gao E., npj Computational Materials, 6[1] 2020, 13, under Creative Commons licence, with original captions removed.

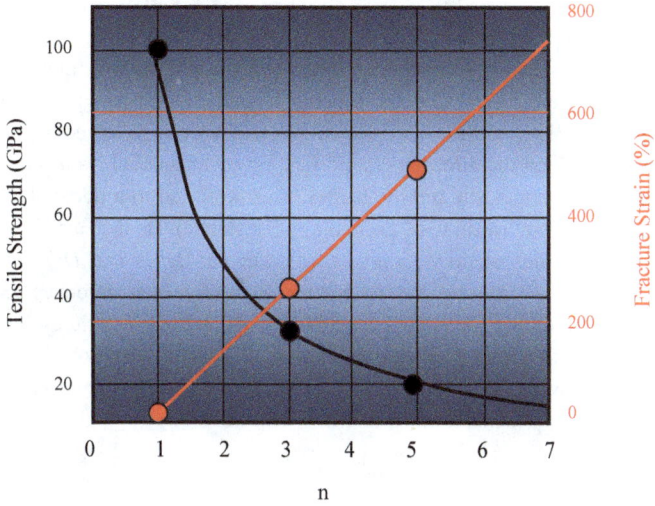

Figure 18. The effect of n upon the tensile strength and fracture strain of self-folded graphene-based materials. Points: simulation results, lines: theoretical predictions

About the Author

Dr. Fisher has wide knowledge and experience of the fields of engineering, metallurgy and solid-state physics, beginning with work at Rolls-Royce Aero Engines on turbine-blade research, related to the Concord supersonic passenger-aircraft project, which led to a BSc degree (1971) from the University of Wales. This was followed by theoretical and experimental work on the directional solidification of eutectic alloys having the ultimate aim of developing composite turbine blades. This work led to a doctoral degree (1978) from the Swiss Federal Institute of Technology (Lausanne). He then acted for many years as an editor of various academic journals, in particular *Defect and Diffusion Forum*. In recent years he has specialized in writing monographs which introduce readers to the most rapidly developing ideas in the fields of engineering, metallurgy and solid-state physics. He is a co-author of the widely-cited student textbook, *Fundamentals of Solidification*, now in its entirely-revised 5th edition. Google Scholar credits him with 9348 citations and a lifetime h-index of 13.

References

[1] Valiev R.Z., Alexandrov, I.V., Zhu Y.T., Lowe T.C., Journal of Materials Research, 17[1] 2002, 5-8. https://doi.org/10.1557/JMR.2002.0002

[2] Valiev R.Z., Sergueeva A.V., Mukherjee A.K., Scripta Materialia, 49, 2003, 669-674. https://doi.org/10.1016/S1359-6462(03)00395-6

[3] Alexandrov I.V., Materials Science and Engineering A, 387-389[1-2] 2004, 772-776. https://doi.org/10.1016/j.msea.2004.01.116

[4] Kumar P., Kawasaki M., Langdon T.G., Journal of Materials Science, 51[1] 2016, 7-18. https://doi.org/10.1007/s10853-015-9143-5

[5] Balasubramanian N., Langdon T.G., Metallurgical and Materials Transactions A, 47[12] 2016, 5827-5838. https://doi.org/10.1007/s11661-016-3499-2

[6] Alexandrov I.V., Chembarisova R.G., Materials Science Forum, 633-634, 2010, 231-248. https://doi.org/10.4028/www.scientific.net/MSF.633-634.231

[7] Liu X., Zhuang L., Zhao Y., Materials, 13, 2020, 5171. https://doi.org/10.3390/ma13225171

[8] Savarabadi M.M., Faraji G., Zalnezhad E., Journal of Alloys and Compounds, 785, 2019, 163-168. https://doi.org/10.1016/j.jallcom.2019.01.149

[9] Li G., Morinaka S., Kawabata M., Ma C., Ameyama K., Procedia Manufacturing 15, 2018, 1641-1648. https://doi.org/10.1016/j.promfg.2018.07.292

[10] Alawadhi M.Y., Sabbaghianrad S., Huang Y., Langdon T.G., Materials Science and Engineering: A, 802, 2021, 140546. https://doi.org/10.1016/j.msea.2020.140546

[11] Hamid M., Jamalian M., De Vincentis N., Buck Q., Field D.P., Journal of Engineering Materials and Technology, Transactions of the ASME, 144[1] 2022, 011013. https://doi.org/10.1115/1.4051901

[12] Peng S., Wei Y., Gao H., Proceedings of the National Academy of Sciences, 117[10] 2020, 5204-5209. https://doi.org/10.1073/pnas.1914615117

[13] Mao Q., Zhang Y., Liu J., Zhao Y., Nano Letters, 21, 2021, 3191-3197. https://doi.org/10.1021/acs.nanolett.1c00451

[14] Han Q., Li J., Yi X., Journal of the Mechanics and Physics of Solids, 173, 2023, 105200. https://doi.org/10.1016/j.jmps.2023.105200

[15] Wang X., Jiang L., Cooper C., Yu K., Zhang D., Rupert T.J., Mahajan S., Beyerlein, I.J., Lavernia E.J., Schoenung J.M., Acta Materialia, 195, 2020, 468-481. https://doi.org/10.1016/j.actamat.2020.05.021

[16] Wu X., Yuan F., Yang M., Jiang P., Zhang C., Chen L., Wei Y., Ma E., Scientific Reports, 5, 2015, 11728. https://doi.org/10.1038/srep11728

[17] Dai P., Xu W., Tang D., Journal of Physics - Conference Series, 240, 2010, 012149. https://doi.org/10.1088/1742-6596/240/1/012149

[18] Sitdikov V., Alexandrov I., Reviews on Advanced Materials Science, 31, 2012, 85-89.

[19] Alexandrov I., Chembarisova R., Sitdikov V., Kazyhanov V., Materials Science and Engineering A, 493[1-2] 2008, 170-175. https://doi.org/10.1016/j.msea.2007.11.073

[20] Alexandrov I.V., Chembarisova R.G., International Journal of Microstructure and Materials Properties, 7[2-3] 2012, 187-204. https://doi.org/10.1504/IJMMP.2012.047499

[21] Wu X., Yang M., Yuan F., Wu G., Wei Y., Huang X., Zhu Y., Proceedings of the National Academy of Sciences, 112[47] 2015, 14501-14505. https://doi.org/10.1073/pnas.1517193112

[22] Zhilyaev A.P., Huang Y., Cabrera J.M. Langdon T.G., Defect and Diffusion Forum, 385, 2018, 284-289. https://doi.org/10.4028/www.scientific.net/DDF.385.284

[23] Wojtas D., Maj L., Jarzebska A., Wierzbanowski K., Chulist R., Kawalko J., Trelka A., Sztwiertnia K., Materials Letters, 325, 2022, 132831. https://doi.org/10.1016/j.matlet.2022.132831

[24] Catherine L.D.K., Hamid D.B.A., IOP Conference Series - Materials Science and Engineering, 429, 2018, 012014. https://doi.org/10.1088/1757-899X/429/1/012014

[25] Wang Z., Cui X., Chen L., Zhang Y., Ding H., Zhang Y., Gao N., Cong G., An Q., Wang S., Chen J., Geng L., Huang L., Materials Science and Engineering A, 891, 2024, 14592. https://doi.org/10.1016/j.msea.2023.145926

[26] Miao K.S., Xia Y.P., Li D.Y., Wu H., Fan G.H., IOP Conference Series - Materials Science and Engineering, 1249, 2022, 012065. https://doi.org/10.1088/1757-899X/1249/1/012065

[27] Wei Q., Zhang H.T., Schuster B.E., Ramesh K.T., Valiev R.Z., Kecskes L.J., Dowding R.J., Magness L., Cho K., Acta Materialia, 54[15] 4079-4089. https://doi.org/10.1016/j.actamat.2006.05.005

[28] Xia Y.P., Wu H., Miao K.S., Geng L., Fan G.H., Yu T., Jensen D.J., IOP Conference Series - Materials Science and Engineering, 1249, 2022, 012037. https://doi.org/10.1088/1757-899X/1249/1/012037

[29] Yun Y.L., Pan C., Rui J.X., Cheng J.L., Materials Research Express, 7. 2020, 066510. https://doi.org/10.1088/2053-1591/ab990f

[30] Mungole T., Kumar P., Kawasaki M., Langdon T.G., Journal of Materials Science, 50, 2015, 3549-3561. https://doi.org/10.1007/s10853-015-8915-2

[31] Du J., Liu Y., Zhang Z., Xu C., Gao K., Dai J., Liu F., Materials Today Communications, 37, 2023, 107218. https://doi.org/10.1016/j.mtcomm.2023.107218

[32] Mungole T., Kumar P., Kawasaki M., Langdon T.G., Journal of Materials Research, 29[21] 2014, 2534-2546. https://doi.org/10.1557/jmr.2014.272

[33] Kumar P., Kawasaki M., Langdon T.G., Materials Science Forum, 879, 2017, 1043-1048. https://doi.org/10.4028/www.scientific.net/MSF.879.1043

[34] Muñoz J.A., Huvelle L., Huerta E.M., Cabrera J.M., Materials Research Proceedings, 32, 2023, 330-337.

[35] Li S.H., Zhang J., Han W.Z., Scripta Materialia, 165, 2019, 112-116. https://doi.org/10.1016/j.scriptamat.2019.02.025

[36] Zhao Y., Transactions of Nonferrous Metals Society of China, 31[5] 2021, 1205-1216.

[37] Alexandrov I.V., Zhilina M.V., Scherbakov A.V., Bonarski J.T., Materials Science and Engineering A, 410-411, 2005, 332-336. https://doi.org/10.1016/S1003-6326(21)65572-3

[38] Alexandrov I.V., Chembarisova R.G., Sitdikov V.D., Materials Science and Engineering A, 463[1-2] 2007, 27-35. https://doi.org/10.1016/j.msea.2005.08.110

[39] Alexandrov I.V., Chembarisova R.G., Zainullina L.I., Wei K.X., Wei W., Hu J., Journal of Materials Research, 31[24] 2016, 3850-3859. https://doi.org/10.1016/j.msea.2006.07.155

[40] Alexandrov I.V., Chembarisova R.G., Reviews on Advanced Materials Science, 16, 2007, 51-72. https://doi.org/10.1557/jmr.2016.451

[41] Alexandrov I.V., Chembarisova R.G., Reviews on Advanced Materials Science, 31, 2012, 91-99.

[42] Wang Q., Yang Y., Jiang H., Liu C.T., Ruan H.H., Lu J., Scientific Reports, 4, 2014, 4757. https://doi.org/10.1038/srep04757

[43] Meng A., Nie J., Wei K., Kang H., Liu Z., Zhao Y., Vacuum, 167, 2019, 329-335. https://doi.org/10.1016/j.vacuum.2019.06.027

[44] Tian X., Zhao Y., Gu T., Guo Y., Xu F., Hou H., Materials Science and Engineering A, 849, 2022, 143485. https://doi.org/10.1016/j.msea.2022.143485

[45] Bae S.H., Jung K.H., Shin Y.C., Yoon D.J., Kawasaki M., Materials Characterization, 112, 2016, 105-112. https://doi.org/10.1016/j.matchar.2015.12.009

[46] Pereira T.S., Chung C.W., Ding R., Chiu Y.L., IOP Conference Series - Materials Science and Engineering, 4, 2009, 012022. https://doi.org/10.1088/1757-899X/4/1/012022

[47] Dang C., Wang J., Wang J., Yu D., Zheng W., Xu C., Wang Z., Feng L., Chen X., Pan F., Vacuum, 215, 2023, 112275. https://doi.org/10.1016/j.vacuum.2023.112275

[48] Sun Y., Li W., Shi X., Tian L., Materials Research Express, 7, 2020, 116520. https://doi.org/10.1088/2053-1591/abc911

[49] Rai R.K., Sahu J.K., Materials Letters, 220, 2018, 90-93. https://doi.org/10.1016/j.matlet.2018.02.128

[50] Yapici G.G., Materials Letters, 279, 2020, 128443. https://doi.org/10.1016/j.matlet.2020.128443

[51] Zafari A., Xia K., Materials Research Letters, 6[11] 2018, 627-633. https://doi.org/10.1080/21663831.2018.1525773

[52] Cvijović-Alagić I., Cvijović Z., Maletaškić Z., Rakin M., Materials Science and Engineering A, 736, 2018, 175-192. https://doi.org/10.1016/j.msea.2018.08.094

[53] Yao Z., Jia X., Yu J., Yang M., Huang C., Yang Z., Wang C., Yang T., Wang S., Shi R., Wei J., Liu X., Materials and Design, 225, 2023, 111559. https://doi.org/10.1016/j.matdes.2022.111559

[54] Li M.F., Wang P.W., Malomo B., Yang L., International Journal of Plasticity, 169, 2023, 103734. https://doi.org/10.1016/j.ijplas.2023.103734

[55] Kuramoto S., Furuta T., Nagasako N., Horita Z., Applied Physics Letters, 95[21] 2009, 211901. https://doi.org/10.1063/1.3266832

[56] Benito J.A., Tejedor R., Rodríguez-Baracaldo R., Cabrera J.M., Prado J.M., Materials Science Forum, 633-634, 2010, 197-203. https://doi.org/10.4028/www.scientific.net/MSF.633-634.197

[57] Branagan D.J., Frerichs A.E., Meacham B.E., Cheng S., Sergueeva A.V., SAE Technical Papers, 2016, 01-0357.

[58] Tsuji N., Journal of Physics - Conference Series, 165, 2009, 012010. https://doi.org/10.1088/1742-6596/165/1/012010

[59] Terada D., Ikeda G., Park M., Shibata A., Tsuji N., IOP Conference Series - Materials Science and Engineering, 219, 2017, 012008. https://doi.org/10.1088/1757-899X/219/1/012008

[60] Zhang L., Jiang X., Wang Y., Chen Q., Chen Z., Zhan Y., Huang T., Wu G., IOP Conference Series - Materials Science and Engineering, 219, 2017, 012052. https://doi.org/10.1088/1757-899X/219/1/012052

[61] Hohenwarter A., Kapp M.W., Völker B., Renk O., Pippan R., IOP Conference Series - Materials Science and Engineering, 219, 2017, 012003. https://doi.org/10.1088/1757-899X/219/1/012003

[62] Rosenstock D., Banik J., Gerber T., Myslowicki S., IOP Conference Series - Materials Science and Engineering, 651, 2019, 012040. https://doi.org/10.1088/1757-899X/651/1/012040

[63] Rinaldo H., Januar M., Sinaga M.M., IOP Conference Series - Materials Science and Engineering, 541, 2019, 012019. https://doi.org/10.1088/1757-899X/541/1/012019

[64] Wang D., Huang L., Wang K., Wang X., Wang X., Wang W., Hao G., Journal of Materials Research and Technology, 24, 2023, 3746-3758. https://doi.org/10.1016/j.jmrt.2023.04.080

[65] Song M., Sun C., Fan Z., Chen Y., Zhu R., Yu K.Y., Hartwig K.T., Wang H., Zhang X., Acta Materialia, 112, 2016, 361-377. https://doi.org/10.1016/j.actamat.2016.04.031

[66] Shang Z., Ding J., Fan C., Song M., Li J., Li Q., Xue S., Hartwig K.T., Zhang X., Acta Materialia, 169, 2019, 209-224. https://doi.org/10.1016/j.actamat.2019.02.043

[67] Li T., Hu Q., Ma G., Liu W., Wu G., Mao X., Journal of Physics - Conference Series, 2635, 2023, 012011. https://doi.org/10.1088/1742-6596/2635/1/012011

[68] Wang Y.H., Kang J.M., Peng Y., Zhang H.W., Wang T.S., Huang X., IOP Conference Series - Materials Science and Engineering, 219, 2017, 012043. https://doi.org/10.1088/1757-899X/219/1/012043

[69] Arora H.S., Ayyagari A., Saini J., Selvam K., Riyadh S., Pole M., Grewal H.S., Mukherjee S., Scientific Reports, 9. 2019, 1972. https://doi.org/10.1038/s41598-019-38707-3

[70] Saini J., Arora H.S., Grewal H.S., Perumal G., Ayyagari A., Salloom R., Mukherjee S., Steel Research International, 90[5] 2019, 1800554. https://doi.org/10.1002/srin.201800554

[71] Zheng Z.J., Liu J.W., Gao Y., Materials Science and Engineering A, 680, 2017, 426-432. https://doi.org/10.1016/j.msea.2016.11.004

[72] Kong H.J., Yang T., Chen R., Yue S.Q., Zhang T.L., Cao B.X., Wang C., Liu W.H., Luan J.H., Jiao Z.B., Zhou B.W., Meng L.G., Wang A., Liu C.T., Scripta Materialia, 186, 2020, 213-218. https://doi.org/10.1016/j.scriptamat.2020.05.008

[73] Wang B., Niu M., Wang W., Jiang T., Luan J., Yang K., Acta Metallurgica Sinica, 59[5] 2023 636-646.

[74] Hung C.Y., Shimokawa T., Bai Y., Tsuji N., Murayama M., Scientific Reports, 11[1] 2021, 19298. https://doi.org/10.1038/s41598-021-87811-w

[75] Kong H., Jiao Z., Lu J., Liu C.T., Science China Materials, 64[7] 2021, 1580-1597. https://doi.org/10.1007/s40843-020-1595-2

[76] Dobatkin S., Zrnik J., Nikulin S., Kovarik T., Journal of Physics - Conference Series, 240, 2010, 012127. https://doi.org/10.1088/1742-6596/240/1/012127

[77] Xu Z., Shen X., Allam T., Song W., Bleck W., Journal of Materials Research and Technology, 17, 2022, 2601-2613. https://doi.org/10.1016/j.jmrt.2022.02.008

[78] Rezazadeh V., Hoefnagels J.P.M., Geers M.G.D., Peerlings R.H.J., European Journal of Mechanics A, 2023, 105152. https://doi.org/10.1016/j.euromechsol.2023.105152

[79] He Q., Wang Y.F., Wang M.S., Guo F.J., Wen Y., Huang C.X., Materials Science and Engineering A, 780, 2020, 139146. https://doi.org/10.1016/j.msea.2020.139146

[80] Wang Y., Ding Z., Gao Z., Wang X., Journal of Materials Science, 57, 2022, 15530-15548. https://doi.org/10.1007/s10853-022-07562-5

[81] Pan Q., Guo S., Cui F., Jing L., Lu L., Nanomaterials, 11, 2021, 2613. https://doi.org/10.3390/nano11102613

[82] Shen Z., Qiu N., Zhang Y., Zuo X., Metallurgical and Materials Transactions A, 54, 2023, 688-706. https://doi.org/10.1007/s11661-022-06917-6

[83] Jiang W., Zhu Y., Zhao Y., Frontiers in Materials, 8, 2021, 792359.

[84] Ma E., Wu X., Nature Communications, 10, 2019, 5623. https://doi.org/10.1038/s41467-019-13311-1

[85] Dasari S., Sharma A., Jiang C., Gwalani B., Lin W.C., Lo K.C., Gorsse S., Yeh A.C., Srinivasan S.G., Banerjee R., Proceedings of the National Academy of Sciences of the USA, 120[23] 2023, e2211787120. https://doi.org/10.1073/pnas.2211787120

[86] Sharma A., Dasari S., Ingale T., Jiang C., Gwalani B., Srinivasan S.G., Banerjee R., Acta Materialia, 258, 2023, 119248. https://doi.org/10.1016/j.actamat.2023.119248

[87] Du X.H., Huo X.F., Chang H.T., Li W.P., Duan G.S., Huang J.C., Wu B.L., Zou N.F., Zhang L., Materials Research Express, 7, 2020, 034001. https://doi.org/10.1088/2053-1591/ab7a64

[88] Bahadur F., Jain R., Biswas K., Gurao N.P., International Journal of Fatigue, 155, 2022, 106545. https://doi.org/10.1016/j.ijfatigue.2021.106545

[89] Xu W.W., Xiong Z.Y., Li Z.N., Gao X., Li W., Yang T., Li X.Q., Vitos L., Liu C.T., International Journal of Plasticity, 158, 2022, 10343. https://doi.org/10.1016/j.ijplas.2022.103439

[90] Podolskiy A.V., Shapovalov Y.O., Tabachnikova E.D., Tortika A.S., Tikhonovsky M.A., Joni B., Odor E., Ungar T., Maier S., Rentenberger C., Zehetbauer M.J., Schafler E., Advanced Engineering Materials, 22[1] 2020, 1900752. https://doi.org/10.1002/adem.201900752

[91] Peng Y., Gong J., Christiansen T.L., Somers M.A.J., Materials Letters, 283, 2021, 128896. https://doi.org/10.1016/j.matlet.2020.128896

[92] Garg M., Grewal H.S., Sharma R.K., Gwalani B., Arora H.S., Journal of Alloys and Compounds, 933, 2023, 167750. https://doi.org/10.1016/j.jallcom.2022.167750

[93] Chang H., Zhanga T.W., Ma S.G., Zhao D., Xiong R.L., Wang T., Li Z.Q., Wang, Z.H., Materials and Design, 197, 2021, 109202. https://doi.org/10.1016/j.matdes.2020.109202

[94] Pan Q., Zhang L., Feng R., Lu Q., An K., Chuang A.C., Poplawsky J.D., Liaw P.K., Lu L., Science, 374, 2021, 984-989. https://doi.org/10.1126/science.abj8114

[95] Qin S., Yang M., Jiang P., Wang J., Wu X., Zhou H., Yuan F., Journal of Materials Science and Technology, 153, 2023, 166-180. https://doi.org/10.1016/j.jmst.2023.01.014

[96] Qin S., Yang M., Jiang p., Wang J., Wu X., Zhou H., Yuan F., Acta Materialia, 230, 2022, 117847. https://doi.org/10.1016/j.actamat.2022.117847

[97] Jiang W., Wang H., Li Z., Zhao Y., Journal of Materials Science and Technology, 144, 2023, 128-133. https://doi.org/10.1016/j.jmst.2022.10.024

[98] Tsai C.W., Tsai M.H., Tsai K.Y., Chang S.Y., Yeh J.W., Materials Science and Technology, 31, 2015, 1178-1183. https://doi.org/10.1179/1743284714Y.0000000754

[99] Jiang W., Yuan S., Cao Y., Zhang Y., Zhao Y., Acta Materialia, 213, 2021, 116982. https://doi.org/10.1016/j.actamat.2021.116982

[100] Wang L., Jiao Y., Liu R., Wang D., Yu Z., Xi Y., Zhang K., Xu S., Liu H., Wen L., Xiao X., Zhang W., Ji J., Journal of Metals, 76[1] 2024, 353-360. https://doi.org/10.1007/s11837-023-06217-3

[101] Xu H., Zhang M., Zhang G., Li G., International Journal of Refractory Metals and Hard Materials, 118, 2024, 106499. https://doi.org/10.1016/j.ijrmhm.2023.106499

[102] Gao X., Jiang W., Lu Y., Ding Z., Liu J., Liu W., Sha G., Wang Y., Li T., Chang I.T.H., Zhao Y., Journal of Materials Science and Technology, 154, 2023, 166-177. https://doi.org/10.1016/j.jmst.2023.01.023

[103] Esawi A.M.K., Morsi K., Salama I., Saleeb H., TMS Annual Meeting, 1, 2012, 545-552. https://doi.org/10.1002/9781118356074.ch70

[104] Bharath L., Reddy M.S., Girisha H.N., Balakumar G., IOP Conference Series - Materials Science and Engineering, 1055, 2021, 012117. https://doi.org/10.1088/1757-899X/1055/1/012117

[105] Dar S.M., Zhao Y., Kai X., Xu Z., Materials Characterization, 201, 2023, 112913. https://doi.org/10.1016/j.matchar.2023.112913

[106] Zou B., Wang L., Zhang Y., Liu Y., Ouyang Q., Jin S., Zhang D., Yan W., Li Z., Materials Research Letters, 11[5] 2023, 360-366. https://doi.org/10.1080/21663831.2022.2153630

[107] Xue Y., Hao Q., Li N., Wang X., Yin C., Zhang H., Materials Research Express, 8, 2021, 056519. https://doi.org/10.1088/2053-1591/ac0264

[108] Wei X., Tao J., Hu Y., Liu Y., Bao R., Li F., Fang D., Li C., Yi J., Materials Science and Engineering: A, 816, 2021, 141248. https://doi.org/10.1016/j.msea.2021.141248

[109] Wang D., Yan A., Liu Y., Wu Z., Gan X., Li F., Tao J., Li C., Yi J., Nanomaterials, 12[15] 2022, 2548. https://doi.org/10.3390/nano12152548

[110] Wu S., Liu Y., Yu J., Zhao Q., Tao J., Wu Z., Zhang J., Fan Y., Liu Y., Li C., Yi J., Journal of Materials Research and Technology, 23, 2023, 5066-5081. https://doi.org/10.1016/j.jmrt.2023.02.136

[111] Yan A., Jiang H., Yu J., Zhao Q., Wu Z., Tao J., Li C., Yi J., Liu Y., Materials Science and Engineering: A, 867, 2023, 144500. https://doi.org/10.1016/j.msea.2022.144500

[112] Huang S., Wu H., Zhao Z., Zhu H., Xie Z., Materials Science and Engineering A, 865, 2023, 144319. https://doi.org/10.1016/j.msea.2022.144319

[113] Mungole T., Mansoor B., Ayoub G., Field D.P., Applied Physics Letters, 113[10] 2018, 101902. https://doi.org/10.1063/1.5041344

[114] Pan D., Zhang X., Hou X., Han Y., Chu M., Chen B., Jia L., Kondoh K., Li S., Materials Science and Engineering A, 799, 2021, 140137. https://doi.org/10.1016/j.msea.2020.140137

[115] Xiao Y,, Song C., Liu Z., Liu L., Zhou H., Wang D., Yang Y., International Journal of Extreme Manufacturing, in press.

[116] Yang Y., Liu M., Zhou S., Ren W., Zhou Q., Shi L., Journal of Alloys and Compounds, 871, 2021, 159535. https://doi.org/10.1016/j.jallcom.2021.159535

[117] Zhong Z., Zhang B., Ye J., Ren Y., Zhang J., Ye F., Materials Science and Engineering A, 891, 2024, 145996. https://doi.org/10.1016/j.msea.2023.145996

[118] Lu Y., Su S., Zhang S., Huang Y., Qin Z., Lu X., Chen W., Acta Materialia, 206, 2021, 116632. https://doi.org/10.1016/j.actamat.2021.116632

[119] Jia X., Liu Z., Gao E., npj Computational Materials, 6[1] 2020, 13.

www.ingramcontent.com/pod-product-compliance
Lightning Source LLC
Chambersburg PA
CBHW071713210326
41597CB00017B/2466